THE REFERENCE SHELF

The books in this series contain reprints of articles, excerpts from books, and addresses on current issues and social trends in the United States and other countries. There are six separately bound numbers in each volume, all of which are generally published in the same calendar year. One number is a collection of recent speeches; each of the others is devoted to a single subject and gives background information and discussion from various points of view, concluding with a comprehensive bibliography that contains books and pamphlets and abstracts of additional articles on the subject. Books in the series may be purchased individually or on subscription.

Library of Congress Cataloging-in-Publication Data

Main entry under title:

Our future in space   /   edited by Steven Anzovin.
    p. cm. — (The Reference shelf ; v. 63, no. 2)
   Includes bibliographical references.
   ISBN 0-8242-0812-9
   1. Astronautics—Popular works.   I. Anzovin, Steven.   II. Series.
TL793.087   1991
  629.4—dc20
                                           91-8280
                                            CIP

Cover: Space walker James van Hoften launching a Syncom satellite from Space Ship Discovery.
Photo: AP/Wide World Photos

Printed in the United States of America

CONTENTS

PREFACE

It's a short step, merely sixty years, from Robert Goddard's pioneering rocket experiments in the New Mexico desert in the 1920s and 30s to the ear-splitting thunder of a space shuttle launch at Cape Canaveral. Goddard, developer of the basic liquid fuel engine that is used in most rockets today, could hold his first rocket in his arms. Today's heavy launch vehicles are taller than most buildings, generate millions of pounds of thrust, and can loft payloads of tens of thousands of pounds into the earth's orbit. Goddard's experiments were modest affairs; often, he jealously guarded his research by doing most of the work alone. In 1990, however, the American space program consumed billions of dollars, employed tens of thousands of engineers, workers, and managers, and had one of the highest profiles of any arm of the government.

For all its size and complexity, the space program today is in trouble. In an age of fiscal restraint, when competition with the Soviet Union has given way to cooperation, it is not clear what we are trying to do in space, or even why we are there. Disasters and delays have robbed NASA (National Aeronautics and Space Administration), the government agency charged with managing the national space effort, of much of its glamour. Recent presidents have been unable to articulate a space policy that captures the public imagination the way John Kennedy did in 1960, when he called for a manned landing on the moon within two decades. And public enthusiasm for expensive space exploration wavers as the bills for the savings and loan crisis and other billion-dollar boondoggles come due. Still, no one expects the United States to give up manned spaceflight. In the national psyche, it touches an irrational but deep yearning for adventure. What is required is a space policy that fulfills this need while setting realistic, affordable, profitable goals.

This compilation of articles, essays, and speeches treats several important aspects of the subject. The first section of articles looks back at the high and low points of America's space program to date—the manned landing on the moon and the explosion of the space shuttle Challenger in 1986. In Section II, NASA's problems are discussed in more detail. Is the agency no longer able to meet its mandate to explore space, or is it merely suffering a run of bad luck? The articles in Section III highlight some of the space program's recent successes, and discuss how the commercialization of space will broaden the base of the nation's space capability. The articles in the final section outline long-range perspectives on our future in space. Missions to the moon, Mars, and beyond are all in the planning stages—but without an unprecedented level of international cooperation, and an understanding of our own human flaws and limitations, we may never get there.

The editor wishes to thank the authors and publishers who kindly granted permission to reprint the material in this collection. Special thanks are due to Diane Podell of the B. Davis Schwartz Memorial Library, C.W. Post Center, Long Island University.

STEVEN ANZOVIN

November 1990

I. APOLLO AND CHALLENGER

EDITOR'S INTRODUCTION

American space policy is largely the product of two crucial events: the spectacular success of the Apollo moon landing in 1969, and the shocking disaster of the space shuttle Challenger explosion in 1986. The Apollo years were NASA's golden age: the agency had an attainable mission that was also romantic, adventurous, and immensely popular. NASA showed it was a "can-do" agency that could do more with less, come in ahead of schedule, and still deliver the goods. America showed the world, and in particular the Soviets, that it had what Tom Wolfe called "the right stuff". The men of the Apollo missions joined the pantheon of heroes. They were the last great pioneers who had crossed the final frontier. The first article, "I'm at the Foot of the Ladder," is Harry Hurt III's tribute to the Apollo missions and its astronauts. In it he recalls the first lunar landing.

If Apollo was NASA's greatest triumph, the Challenger disaster, in which seven astronauts were killed when the shuttle blew up shortly after liftoff, was its greatest tragedy. In the five years since the first shuttle flight, the missions had become commonplace. Challenger changed all that. The disaster marked the end of our innocence and the beginning of our disillusionment with America's space program. Initially the accident was believed to have been caused by an undetermined but probably purely technical failure, but it was eventually revealed that it was in fact a structural defect in the O-ring sealing the casing joints of one of the Shuttle's solid rocket boosters. Icy temperatures the night before the launch had caused the rubber ring to crack, allowing burning gases to penetrate the booster casing after liftoff and ignite the main fuel tank. In "An Outsider's Inside View of the Challenger Inquiry," Nobel Prize–winning physicist Richard Feynman, who was the first to publicly espouse the O-Ring theory, offers his own assessment of NASA and the presidential commission created to investigate the Challenger accident. In Feynman's view, Challenger was a disaster waiting to happen. He contends that it was those very same qualities that had been such

7

an asset to NASA during the Apollo years: arrogance, defiance of the laws of chance and adherence to the "can do" attitude, that conspired against them and contributed to the Challenger accident.

However, the blame is not all NASA's, as the final two selections illustrate. "Two Minutes" by Tom Bancroft, reprinted from *Financial World*, and "Whistle-blower" by Toni Chiu, reprinted from *Life*, both focus on the inner dynamics at Morton Thiokol, the NASA contractor that manufactured the solid-fuel boosters. The articles document hazardous working conditions, poor management-employee relations, and crucial communication breakdowns at Morton Thiokol. Engineers at Morton Thiokol had argued against the launch but were overruled by managers eager to get on with the flight.

I'M AT THE FOOT OF THE LADDER[1]

The Public Affairs Officer anxiously watched the ticking of the countdown clock.

"We're coming up on the 60-second mark. We're at T minus 60 seconds and counting. We are GO for a mission to the Moon at this time. . . . "

At T minus 12 seconds, the swing arms bracing the Saturn 5 started to pull away, and the ignition sequence began.

At T minus 8.9 seconds, the first fireballs burst out the end of the trench. But because the booster was still locked tightly to the launch pad by four giant metal hold-down clamps, it did not move. Instead, it continued steadily building thrust.

"Three!"

"Two!"

"Launch commit."

"Zero!"

At that instant, all four hold-down clamps flopped back simultaneously. Seeing the clamps release, the PAO cried, "Liftoff! We have a liftoff!"

[1]Adapted from *For All Mankind* copyright © 1988 by Harry Hurt III and reprinted with permission of Atlantic Monthly Press. A Morgan Entrekin Book.

The three crewmen atop the 360-foot-tall Saturn 5 could not share this view because they had no television monitors on board. Insulated in their "Moon cocoons," they couldn't even hear the rocket's full roar. But all three could feel the bone-jarring quake of liftoff and the gut-wrenching low-frequency shock waves of the booster. And as *Apollo 16*'s Ken Mattingly reports, "It feels just like it sounds."

Apollo 17's Jack Schmitt adds that the sound of the launch made it feel violent. "The low-frequency vibration is the unusual thing that you'll never get off the television. Some of the films of the crowd watching the launch show them breathing very deeply. It's that low-frequency sound that's moving your innards, your tie and your blouse. That's what makes it an emotional experience even for the viewers."

The trans-lunar injection (TLI) usually began about two-and-a-half hours after lift off. The three crewmen were usually making their second orbit over the Pacific Ocean. Like staging and every other engine burn during the mission, TLI was always announced by a professionally matter-of-fact exchange between Houston and the astronauts.

"You are GO for TLI," rasped the CAPCOM.

"Roger," replied the command module pilot. "Thank you."

As *Apollo 11*'s Mike Collins observes, "There ought to be more to it."

The three astronauts realized that TLI was not just another engine burn, but the one that would enable them to set sail for the Moon.

The TLI burn was often, in the words of *Apollo 11*'s Buzz Aldrin, "just a tiny bit rattly," but it was nowhere near as bone-jarring as liftoff. The astronauts felt only about one G from the thrust of the S-IVB third stage, and as the burn progressed, they were treated to some breathtaking visuals.

"I just wish I would have had a camera at that point," exclaims Jim Irwin of *Apollo 15*. "I looked out my window, and there were all the Hawaiian islands. It was a clear morning on the Pacific and you could see the volcanos of Mauna Loa and Mauna Kea. We're being lifted up majestically, accelerating to 36,000 miles per hour and you see the islands fade, get smaller and smaller."

The TLI burn lasted for 5 minutes and 47 seconds. Then the third-stage engine shut down, the rattling stopped, and the long trans-lunar coast began.

The astronauts were slightly disoriented by the lack of milestones to mark their progress through cislunar space. As *Apollo
12*'s Al Bean observes, "One of the things different about a lunar
trip is that first you leave the launch pad, then you leave Earth
orbit, then about a couple of days later, after just floating along
and watching the Earth get smaller, all of a sudden you're at the
Moon. The lack of way points made it seem a little magical or
mystical. All of a sudden, you just showed up."

Lunar orbit insertion (LOI) was a braking maneuver, the reverse of the trans-lunar injection that blasted the spacecraft out
of Earth orbit. LOI was much trickier and much riskier, however,
because it always took place as the astronauts rounded the back
side of the Moon. As a result, the crew had to get themselves into
lunar orbit without the guidance of the ground.

Invariably, there was almost total silence inside the command
module as the three crewmen barreled toward the Moon. At the
moment of losing radio contact, they were supposed to be a good
300 nautical miles above and to the left of the lunar hemisphere
visible from Earth. But due to the increasing pull of lunar gravity
and the trajectory of the spacecraft, they appeared to be plunging
in a suicide dive toward the lunar surface.

"You're coming in awful fast," recalls *Apollo 12* command
module pilot Dick Gordon, "and the Moon is growing—it's visibly getting larger by the minute—and it looks like you're going
to run into the middle of the damn thing. You know intellectually
that you're not gonna hit it, but what your eyes see and what your
intellect tells you are two different things."

"You're drifting in," says Stu Roosa, "and that old Moon is
growing magnificently fast, just filling up the window—pale
brown, pale gray, turning black. Then you drift into its shadow
and turn around the back side. Soon you're in total darkness.

"You make the LOI burn in the dark, and then without any
warning you pop into the light and look out the window for your
first good close-up view of the Moon. At that point, we were still
about 120 miles above the surface, but it looked so close you
thought you could just reach out and touch it."

At first sight, all the astronauts found the lunar landscape for-
bidding. *Apollo 11*'s Mike Collins recalls, "The Sun was hitting the
Moon at a very, very shallow angle, and the craters were throwing
very long shadows. The area looked so chopped up and so sharp,
I thought, 'My God, there's no way they're going to find a place

smooth enough to put the spacecraft down in that kind of terrain. There's just no way.'"

By the end of the fourth day into the flight of *Apollo 11*, as the crew embarked on their seventh orbit around the Moon, the anxiety level inside the spacecraft began to climb. The undocking and separation maneuver was now less than half a day away. The first lunar touchdown itself was only fifteen hours away.

Awakened by the crackle of the radio, Armstrong and Aldrin tumbled out of their hammocks to help Collins fix breakfast. Both were starting to feel the pressure. Armstrong had gotten less than six hours of sleep, Aldrin only five. Despite the fact that they were beginning their fifth day in space and were also veterans of *Gemini* flights, they fumbled with their color-coded food packets like nervous rookies.

The spacecraft slipped behind the Moon for the tenth time, breaking contact with Houston. Aldrin crawled through the access tunnel to the Lunar Module *Eagle*, which had been fully pressurized the day before, and powered up the main control panels. Then he floated back to the Command Module *Columbia*, where he and Armstrong doffed their long johns and began putting on their "Moon cocoons."

"I got as far as my liquid-cooled underwear," Aldrin recalls, "and left to make some initial checks in the LM while Neil took my place in the navigation bay to begin changing. There is only room for one man to change clothes, with another standing by to help pull the long crotch-to-shoulder zipper closed. Both Neil and I were determined to put on our suits as perfectly and comfortably as possible—we were going to be in them for quite some time."

The astronauts prepared for undocking during orbit twelve. Entering *Eagle* and closing the hatch behind them, Armstrong and Aldrin pressurized their suits and cross-checked the LM's guidance platform, the rendezvous radar, the descent propulsion system, and the thrusters used for maneuvers.

"*Apollo 11*, Houston," called CAPCOM Charley Duke as the spacecraft went around to the back side of the Moon. "We're GO for undocking." Collins threw the release switch, and a set of explosive bolts popped *Eagle* away from the command module.

Eagle did a slow pirouette in the silence of the lunar void, extending its four spindly legs and lumpy-headed body like a giant insect bursting from its cocoon. A few minutes later, *Columbia* and

Eagle sailed around the front side of the Moon, where Houston impatiently demanded to know the results of the undocking maneuver.

"The *Eagle* has wings!" Collins cried.

Eagle, however, was still 60 miles above the lunar surface and only a few yards away from *Columbia*. To descend to the Sea of Tranquility, the astronauts had to execute three equally crucial and difficult maneuvers.

First, Collins fired the thrusters of the service module for a few quick seconds. That pushed *Columbia* far enough away from *Eagle* for Armstrong and Aldrin to begin the two-step descent procedure.

They knew they could not simply trust their fate to *Eagle*'s automatic guidance system. Although the crew had to rely on the on-board computer to calculate the rapidly changing altitude, attitude, velocity, and fuel consumption figures in the initial stages of the descent, Armstrong would have to assume manual control at 500 feet altitude, perhaps even sooner. As Aldrin remarked before liftoff, "The computer isn't going to dodge boulders for you."

Eagle disappeared behind the Moon, then at Armstrong's command, the descent engine fired and started the feet-first plunge toward the surface. A few minutes later, and due to arrive on the Sea of Tranquility in just 12 minutes, Armstrong and Aldrin were 50,000 feet above the lunar surface with a forward velocity of 3,000 miles per hour. And unbeknowst to them and the ground, they were about to confront a series of last-minute crises.

Four minutes into the descent, *Eagle* rolled over to lock in its landing radar, and the crew found themselves upside down, with "Earth right out our front window," as Aldrin put it.

But just before he lost sight of the lunar surface, Armstrong realized he had a problem. As *Eagle* descended, he had been charting their progress by landmarks, comparing their actual pass over times to those in the flight plan. When a crater called Maskelyne W appeared in his window 2 seconds sooner then expected, he knew something was wrong. Two seconds translated into 2 miles off course.

"Our position checks downrange show us a little long," Armstrong reported.

"Roger, we confirm," replied Charley Duke.

The matter-of-fact exchange belied the gravity of the error. The spacecraft was diving toward the surface 15 miles per hour faster than planned. Mission rules said that if the velocity increased another 8 miles per hour, the astronauts would have to abort the landing. As *Eagle* dropped lower, Houston decided the error would remain in bounds and allowed the landing to continue.

Then the LM's computer flashed an unexpected warning signal.

"Program alarm," rasped Aldrin. "It's a 1202."

Thirty seconds passed while Mission Control sought to identify it. "Give me a readng on that alarm," Aldrin demanded.

The 1202 alarm signified "executive overflow"—the computer was being taxed beyond capacity. The cause, however, turned out to be nothing very esoteric: The rendezvous radar had been left on and the computer was attempting to locate a touchdown site on the Sea of Tranquility at the same time as it was trying to plot a course back to *Columbia*. Control told the crew to ignore the alarm.

A little over eight minutes into the descent, *Eagle* reached 7,000 feet and swung upright with its landing legs down. The braking thrust of the descent engine had reduced the velocity to about 60 miles per hour and the exhaust nozzle now pointed down toward the lunar surface. But according to the mission rules, they had to touch down in less than six minutes or hit the abort handle.

"*Eagle*, Houston. You are GO for landing."

Then at 2,500 feet, *Eagle*'s computer registered a second program alarm.

"1201," barked Aldrin.

Houston determined that this indicated an overload similar to the earlier one and gave the GO to proceed. In the meantime, as *Eagle* passed the 1,000-foot mark, the astronauts appeared to be plummeting toward a large crater full of jagged boulders. Armstrong later admitted, "I was surprised by the size of those boulders. Some were as big as small motor cars. And it seemed we were coming up on them pretty fast. Of course, the clock runs at about triple speed in such a situation."

When Armstrong took over control at 500 feet, he could not decide what to do. He made the LM pitch forward so he could "skim over the top of the boulder field." But this extended the approach trajectory. The spacecraft was now at least four miles off course. And it was rapidly running out of fuel.

"I changed my mind a couple of times again, looking for a parking place," Armstrong recalls. "Something would look good, and then as we got closer, it really wasn't good. Finally we found an area ringed on one side by fairly good-sized craters and on the other side by a boulder field. It was about the size of a big house lot, but it looked satisfactory. I was quite concerned about the fuel level. We had to get on the surface very soon or fire the ascent engine and abort."

At 200 feet, he jerked the LM back into the upright position required for landing.

"Sixty seconds," warned CAPCOM Charley Duke, indicating one minute of descent fuel was left.

"Forty feet," Aldrin reported, "down two and one half."

Eagle drifted into the darkness of its own shadow, and quickly became engulfed in an enormous cloud of dust kicked up by the exhaust of the descent engine.

By this time, Mission Control had started the countdown for an emergency abort, but the astronauts were already in the "dead man's zone." As Aldrin noted later, "If anything had gone wrong, it would probably have been too late to do anything about it before we impacted with the Moon."

"Thirty seconds," CAPCOM Duke reported.

Eagle now had less than half a minute of fuel.

"Forward drift?" Armstrong asked.

"Yes," Aldrin confirmed, adding a reassuring, "Okay . . . "

Duke started to call out another fuel consumption reading, but chief astronaut Deke Slayton, who had taken the adjacent seat at the CAPCOM's console, cut him off.

"Shut up, Charley," Slayton ordered, "and let 'em land."

Moments later, *Eagle* gently plopped down somewhere on the uncharted western edge of the Sea of Tranquility. According to Aldrin, "The landing was so smooth that I had to check the lights from the touch down sensors to make sure the slight bump I felt was indeed the landing."

"CONTACT LIGHT!" Aldrin exulted.

Then, keeping to the official flight plan, he immediately began the post-touchdown checklist.

"Okay," Aldrin noted, "Engine STOP. ACA out of detent."

"Got it," Armstrong confirmed.

"Mode control, both AUTO," Aldrin continued. "Descent engine command override, OFF. Engine arm, OFF. 413 is in."

"We copy you down, *Eagle*," CAPCOM Duke called a few seconds later. There was a pause.

Then Neil Armstrong said:

"Houston, Tranquility Base here. The *Eagle* has landed."

Armstrong and Aldrin expected moonwalk preparations to take about two hours, but they ended up taking twice that long, partly because the exhaust from their backpacks compounded the difficulty of depressurizing the cabin of the LM. Nearly two hours later than anticipated, *Eagle* finally bled off enough cabin pressure for the astronauts to open the hatch.

With Aldrin's help, Armstrong backed through the hatch onto the porch above the ladder. There were now just nine rungs between him and the lunar surface. He activated the television camera (attached to part of the LM) and backed down the ladder, hopping the last three-and-a-half feet to the footpad.

"I'm at the foot of the ladder. . . . The LM footpads are only depressed in the surface about one or two inches. . . . I'm going to step off the LM now. . . ."

At 9.56 p.m. Houston time, Neil Armstrong lifted his right boot off the pad, planted it in the dust of the Sea of Tranquility, and blurted:

"*That's one small step for . . . man, one giant leap for mankind.*"

Armstrong's second, third, and fourth steps were a series of short backward shuffles. Then pausing to steady himself at the ladder, he started kicking at the lunar topsoil to test its depth and texture.

"At the surface it's fine and powdery. I can pick it up loosely with my toe. It adheres in fine layers like powdered charcoal to the sole and sides of my boots. I can go in only a fraction of an inch—maybe an eighth of an inch—but I can see the footprints of my boots and the treads in the fine sandy particles."

When Aldrin joined Armstrong on the surface, he called the sight "magnificent desolation." Aldrin later commented, "I immediately looked down at my feet and became intrigued with the peculiar properties of the lunar dust. On the Moon, dust travels exactly and precisely as it goes, and every grain of it lands very nearly the same distance away."

They collected a number of rock samples and erected the
flag. They then received a phone call from President Richard
Nixon via a special communications link from the White House
to Houston to the Moon. Ceremonials over, Armstrong and Aldrin then set up the Early Apollo Scientific Experiments Package.
Its three main components were a seismometer to measure any
moonquakes, a honeycombed laser retroreflector to let astronomers measure the exact distance between Earth and the Moon,
and a solar wind detector, a sheet of aluminum foil designed to
trap solar particles.

Armstrong claimed that the lunar surface looked like "a nice
place to take a sunbath," and later said, "After landing, we felt
very comfortable in the lunar gravity. It was, in fact, in our view
preferable both to weightlessness and Earth gravity." Armstrong
added his only regret was not being able to spend more time on
the Moon. "We had the problem of the five-year-old boy in the
candy store. There are just too many interesting things to do."

Aldrin, however, later confessed to experiencing some opposite impressions. As he reported in his post-flight memoir, "I
quickly discovered that I felt balanced—comfortably upright—
only when I was tilted slightly forward. I also felt a bit disoriented: On Earth when one looks at the horizon, it appears flat.
On the Moon, so much smaller than the Earth and quite without
high terrain (at least in the Sea of Tranquility), the horizon in all
directions visibly curved down away from us."

Aldrin added that jogging or running on the Moon required
special precautions. "Earth-bound, I would have stopped my run
in just one step—an abrupt halt. I immediately sensed that if I did
this on the Moon, I'd be face down in the lunar dust."

Shortly after the stroke of midnight in Houston, the astronauts were ordered to return to the LM. They later reported that
the first thing they noticed when they removed their helmets inside was a peculiar odor. According to Aldrin, "There was a distinct smell to the lunar material—pungent, like gunpowder or
spent cap pistols. We had carted a fair amount of lunar dust back
inside with us."

The astronauts spent the next three hours housekeeping and
then settled down to sleep. Aldrin tried to curl up on the floor
of the LM, only to discover that he was both too elated and too
cold to sleep. Armstrong rested even more fitfully. Since there
was no room for both to lie on the floor, he slung a hammock be-

tween the ascent engine cover and a support bar. It was decidedly uncomfortable. It didn't help that he was facing up toward the unshuttered portal at the top of the cabin. Earth was suspended above the window like "a big blue eyeball."

Apollo 11's moonwalk lasted only two and a half hours; other crews spent longer on the Moon and spoke more about their experiences. "Do you know what I feel like, Al?" said *Apollo 12*'s Pete Conrad as he bounded over the lunar landscape. "Did you ever see pictures of giraffes running in slow motion? That's just what I feel like." Ed Mitchell on *Apollo 14* bubbled about how he and Alan Shepard were "hopping around like kangaroos."

Apollo 15 carried the first rover vehicle. Dave Scott and Jim Irwin found it provided more than just a physical advantage. "The rover made us feel more at home," Irwin noted, "like we were on Earth, where we could just get into our car and drive."

Jack Schmitt of *Apollo 17* remarked, "I had tried to anticipate what it would be like for many years. But there was no way to anticipate standing in the valley of Taurus-Littrow, seeing this brilliantly illuminated landscape with a brighter Sun than anyone had ever stood in before, with a blacker than black sky, and the mountains rising on either side."

"Lift off from the Moon was probably the greatest anxiety of the whole flight," says *Apollo 16*'s Charley Duke, "primarily because you really had very little to do during the last 15 minutes of the countdown except sit there and think."

The LM had only one ascent engine. There was no back-up. No emergency escape booster. No second chance. The ascent engine had to light on time and keep burning for just enough time to blast the LM into lunar orbit—or else.

Neil Armstrong and Buzz Aldrin deliberately refused to contemplate what might happen if *Eagle*'s ascent engine failed. "That's an unpleasant thing to think about," Armstrong admitted at a pre-launch press conference. "We've chosen not to think about it at the present time."

The crew of *Apollo 12* was nervous, despite the safe return of *Apollo 11*. Commander Pete Conrad tried to ease the mind of his LM pilot, Alan Bean, a rookie.

"Old Al had the fidgy-widgets," Conrad remembers, "and he kept going through his checklist. I said, 'What's the matter, Al? Worried about the engine not lighting?' And he said, 'As a matter of fact, I am.' So I told him he might as well sit back and relax.

If the engine didn't fire, we would become the first permanent monument to the U.S. space program erected on the Moon.

"I'm not sure that gave him the comforting, reassuring words he wanted."

Others felt pangs of regret as they prepared to depart. "I was tired and grimy," recalls *Apollo 14*'s Ed Mitchell. "But I really regretted climbing up that ladder for the last time because I knew I wouldn't have the chance to come back. The feeling was, 'Take a good look. You're not going to see this again.'"

CAPCOM: Tranquility Base, you're cleared for takeoff.

Aldrin: Roger, understand. We're Number 1 on the runway.

Armstrong: Okay, master arm on.

Aldrin: Nine . . . eight . . . seven . . . six . . . five . . . abort stage . . . engine arm ascent . . . proceed!

A split-second later, the ascent engine ignited and the astronauts felt no more than "maybe half of a G or two-thirds of a G" as they began to rise from the lunar surface.

"There was no time to sightsee," Aldrin reported later. "I was concentrating intently on the computers and Neil was studying the attitude indicator. But I looked up long enough to see the flag fall over."

"It's like riding in a fast elevator," remembers Alan Bean. "There's a big bang as you separate, but you never hear the main engine because of the vacuum of space. You hear the valves, but you don't hear them firing. All you hear is a little thump-thump-thump kind of sound.

"As we lifted off I could see these sparkling things being blown off the insulation of the descent stage. It looked like the ripples you see when you drop a rock into a pond, this metallic insulation going out in concentric rings."

A little over seven minutes after liftoff, *Eagle*'s engine shut down and the craft entered a 10-mile by 45-mile elliptical orbit. When it caught up to the orbiting *Columbia*, the two docked securely and Armstrong and Aldrin rejoined Collins.

Although the astronauts inside the LM had risked their lives on the success of the rendezvous, the command module pilot (who remained in orbit) was almost always the happiest of the three when his comrades finally crawled back into the mothership. That was certainly true of *Apollo 12*'s Dick Gordon.

"I've never seen a happier guy in my life," recalls crewmate Al Bean. "He was more happy we got back than we were happy to get back. He was just bananas. He wanted to get us a drink of water. He wanted to fix us some food. He couldn't do enough. It was like returning home to your mother after a twenty-year absence."

All the astronauts eventually had to make the same crucial engine burn to get out of lunar orbit and start the final leg home. This was the trans-Earth injection, or TEI, burn—or as Collins called it, "the get-us-home burn, the save-our-ass burn, the we-don't-want-to-be-a-permanent-Moon-satellite burn."

The TEI burn was never more than three minutes long, but that was powerful enough to make the craft accelerate to lunar escape velocity. As the astronauts were slingshot around the front side of the Moon, they were treated to one of the most thrilling experiences of the entire mission.

"Coming off the Moon is something to behold," recalls *Apollo 14*'s Stu Roosa. "After TEI, you have your windows pointed at the Moon and, oh boy, the way you're hauling out of there. . . ."

Apollo 17's Gene Cernan says the sight of a rapidly shrinking Moon inspired an unexpected nostalgia. "Having lived there for three days, the Moon has a familiarity. You say, 'There's home . . . that's where we left the rover. And there's where we explored . . . and there's the mountains and valleys.' And in a sense, it has been a home for you."

The TEI burn was literally and figuratively the turning point for every Moon landing mission. If the engine failed to light, misfired, or burned too long, the crew would have either remained in lunar orbit or been hurled into space with no hope of being rescued before their oxygen supplies ran out.

"That was the last big hurdle," notes *Apollo 16*'s Charley Duke. "After that, to me, it was home free. We still had reentry to go, but for some reason you just knew you were gonna make it. You might not land right on your splashdown site, but you knew you were gonna be all right."

AN OUTSIDER'S INSIDE VIEW
OF THE CHALLENGER INQUIRY[2]

A few days after the Challenger accident, on a Friday, I got a call from William Graham, who was the acting director of NASA. Mr. Graham had been a student of mine—at Caltech, and also at the Hughes Aircraft Company, where I gave a series of lectures—and thought maybe I would be of some use to the investigation. When I heard it would be in Washington, my immediate reaction was not to do it. I have a principle of not going anywhere near Washington or having anything to do with government.

So I called various friends like Al Hibbs and Dick Davies, trying to find an excuse why I shouldn't accept, but they all said I should. Then I spoke to my wife. "Look," I said. "Anybody could do it. They can get somebody else." "No," said Gweneth. And she explained how she though I would make a unique contribution—in a way that I am modest enough not to describe. Nevertheless, I believed what she said. So I said, "OK. I'll accept."

So on Sunday, as I went to the telephone to call Mr. Graham, I announced to Gweneth, "I'm going to commit suicide for six months. I won't be able to do any work with this physics problem I've been having fun with; I'm going to do nothing but work on the shuttle—for six months." I want you to understand my attitude at the time: I hadn't realized that it would take two years to get the shuttle flying again. I was going to try to work very hard so we could get everything straightened out as quickly as possible.

The next day, Monday, I got a telephone call at 4 pm: "Mr. Feynman, you have been accepted onto the commission"—which by that time was a "Presidential" commission, headed by former Secretary of State William P. Rogers. The first meeting would be in Washington, on Wednesday. So Tuesday, I asked Al Hibbs to get people at the Jet Propulsion Laboratory who knew something about the shuttle project to brief me on it right away. I want to say right now that I got nothing but wonderful cooperation from JPL, and that briefing was fantastic.

[2]Article by physicist Richard Feynman. Reprinted with permission from *Physics Today*. P 26+. F '88 © 1988 by American Institute of Physics.

The first page of the notes I made in the briefing says, "O-rings show scorching in clevis check." That means hot gas had burned through the O-rings on several occasions. Furthermore, they told me that the zinc chromate putty had bubbles, or holes. It turned out that yes, indeed, through those holes the gas came in to erode the O-rings. So already, on the second line of my briefing, I was told what was the matter with the shuttle.

The guys at JPL gave me a lot of other information. They told me about the engines, which are remarkable devices in the sense that the engineering involved is very good. They are way beyond normal. They are the most powerful engines for their weight that have ever been built. NASA was claiming that the engines were in the regular range of engineering, but they're not; the engines had many difficulties that the guys at JPL told me about. (I found out later that the people who worked on the engines always had their fingers crossed on each flight, and the moment they saw the shuttle explode, they were all sure it was the engines. But of course, the TV replay showed a flame coming out of one of the solid rocket boosters.)

Anyway, the point is that I got briefed. And this was done with lots of energy, just like the old days at Los Alamos, one guy after the other: first the rocket, then the engines and so forth. A guy would say, "We don't know about this; Lifer knows about that. Let's get Chuck Lifer in on this." So it was a very intensive briefing, the kind of thing I love, and I sucked up all the information like a sponge. I'm all set to go to Washington, and I go to Washington. (By the way, I took the "red-eye" across the country so I could stay here on Tuesday to learn about the shuttle. But the red-eye I never took again—you're so sleepy when you get there.)

I check into the Holiday Inn early Wednesday morning, I get into a taxi, and read the address of Mr. Rogers's office to the driver. We start off. Mr. Rogers's office was supposed to be near the hotel somewhere—the hotel was located near the Capitol and near everything big—but we go on and on, further and further, into worse and worse territory, until we finally find the address— by interpolation between two numbers. It was an empty lot there, with no number on it.

So now, what to do? I asked the taxi driver to go all the way back over this whole distance. (Meanwhile, my secretary tells me, she got a call from Washington: "Where is he?") Then I noticed that my hotel was right across the street from NASA. Perfect. Right across the street. (In fact, it was also across a different street, on the other corner, from where the commission later had its offices.)

I thought, "What the hell, NASA's right across the street. I'll go to NASA. Somebody there must know where the meeting is." So I went into NASA, up to Mr. Graham's office, and somebody knew. They showed me the room. There, the room was full of people. There were television lights and everything, and all I could do was squash in the back and think, "How the hell am I gonna get to the front where I belong?" I worried about this for a while. Then I overheard a little bit about what they were saying, and it was evidently a different subject!

In the meantime, somebody from Mr. Graham's office had found the location of Mr. Rogers's office by phoning around and came down to get me. I finally made it to Mr. Rogers's law offices a few blocks away, where I met the other commissioners. Over the course of the commission, we all became very good friends. We worked very hard together. This first meeting was the beginning of a very effective commission—with the exception of Mr. Chuck Yeager, who came to one meeting for about a half an hour and then absented himself from the commission in order to be free so he could make criticisms of it.

Well, this first meeting was just a get-together. But Mr. Rogers did discuss the importance of our relationship to the press and how we have to be very careful with the press. "I know Washington," he kept saying. "We have to proceed in an orderly manner and be careful of leaks to the press."

The next meeting we had, on Thursday, was a public meeting—to start things off right with the press. By the way, we arrived at that meeting in limousines. We never got limousines again, but this time we arrived in limousines. I sat in the front seat. The driver says to me, "I understand a lot of very important, famous people are coming to this meeting. . . . "

"Yeah, I s'pose. . . . "

"Well, I collect signatures," he says. "Could you do me a favor . . . "

"Sure," I say.

I'm reaching for my pen when he continues, " . . . and find Mr. Armstrong for me, so I can get his signature?"

There are always greater people.

That meeting was a public briefing. A briefing in a public meeting is almost impossibly inefficient, because other people ask questions, and they're not the questions you want to ask, and you've got to sit through all that, and so on, and so on. It's very ineffective, and I began to learn how boring such things can be. The NASA officials were telling me only a small fraction of all the things I had learned at JPL two days before.

We had all come to the meeting in limousines, and when we came out, some of the limousines were still there. One of the commissioners was a general, General Kutyna, who looked very handsome and very impressive in his uniform. But what impressed me was his request: "Where is the nearest Metro station?" Right away I liked him, and I found out that my judgment in this case was excellent.

That night I wrote out for myself what kinds of questions I thought we should ask and all the things I wanted to study. I laid out the whole business, hoping to see what the rest of the commission wanted to do in our next meeting.

The next day, Friday, was more effective. General Kutyna told us in considerable detail what an accident investigation was like and how it was done, using the Titan missile as an example. I was very impressed with this. I was happy to learn that most of the questions I was going to ask *were* the kinds of questions one should ask, except that the investigation should be done in a much more methodical fashion than I had imagined.

At the end of this discourse, Mr. Rogers, who is not a technical man, said, "Yes, your investigation was a wonderful success, but we can't use those methods on our flight because we can't get as much information as you had on yours." That was patently false, because the shuttle, having people in it, was monitored much more carefully, so we had enormously more information than they had on the Titan. So there wasn't any doubt that we could do it.

In the meeting Mr. Rogers asked each of us how much time we could spend working on the commission. Many of the commissioners were retired, so they could spend 100 percent of their time. I also said I could spend 100 percent; I had everything arranged here at Caltech. (Nobody at Caltech ever said a word to me that I was shirking my work here, and I appreciate that.)

I tried very hard to get something to do. In the meeting I kept explaining that public briefings don't work with me; I have to talk to the technical people directly. Mr. Rogers explained that we were going down to Kennedy Space Center in Florida on the following Thursday. Then we would start our investigation.

Next Thursday? I wanted to get going much quicker than that, and kept explaining that I could work much more efficiently if I went on my own and talked to people directly, and I kept mentioning different things I'd like to do. Then the meeting would be interrupted by a letter coming in for Mr. Rogers, or something. He would read it—during which time various other commission members would whisper to me, "I'd like to work with you if you get a job"—and then Mr. Rogers would look up, apparently forgetting that I had been talking, and call on somebody else.

Finally, I would get the floor again. I would start my stuff again, and there would be another "accident." The meeting stopped while I was still talking, and the last words were by Mr. Armstrong, the vice chairman. He said we wouldn't be doing any of the detailed investigative work. Well, the only thing I'm any good at is detailed work!

I was devastated. I was depressed and very uncomfortable. After the meeting I went up to Mr. Rogers. "Look," I said. "We've got nothing to do for five days!"

He said, "Well, what would you have done if you hadn't been on the commission?"

"I would have gone to Boston to consult for the Thinking Machines Company."

"Well, you go to Boston to consult, and come back in five days."

I couldn't take that. I was wound up like a spring, ready to go to work. I had intended to "commit suicide"—do nothing else but work for the commission—for six months, and I had nothing to do. I was very depressed. I left that meeting feeling terrible.

Soon I thought of something. I called up Mr. Graham, and said, "Listen, Bill, we're not doing anything for *five days!* I want to get started! I want to DO something!"

He says, "Sure! You could go to Johnson, where they take the telemetry; you could go to Marshall, where they make the engines; or you could go to Kennedy."

I didn't want to go to Kennedy, because it would look like I was trying to get information before the rest of the commission did. That was not what I was trying to do; I just wanted to get started. Sally Ride had said she wanted to work with me if I got something to do, and I knew she was at Johnson, so I said I'd go there.

So Graham says, "That's fine, you can do that. I know David Acheson, who's on the commission. He's a good friend of Rogers. I'll call him and see what he thinks." About half an hour later, Mr. Acheson calls me: "I think it's a great idea, but I can't convince Rogers. Rogers refuses to say why he's against it, and I just don't know why I can't convince him that you should get started."

Meanwhile, Mr. Graham thought of a compromise: He would bring people into NASA headquarters, there in Washington, to brief me the next day, on Saturday. But Mr. Rogers called me up and said he didn't want me to do that. He kept explaining that we have to proceed in an orderly manner. I tried to explain how a technical person can talk to another technical person and get information very quickly, and that I wanted to DO something! I complained that we had had several meetings by now, but we hadn't yet discussed who was going to do what, or how to get started on the investigation.

Mr. Rogers said, "Well, do you want me to bother everybody and bring them together again for a meeting on Monday to discuss this?"

I said, "Yes!"

So he dropped the subject. Then he said, "I've heard you don't like your hotel. Let me put you in a good hotel."

I told him everything was fine with the hotel, and that I was perfectly satisfied with it. I just wanted to get to work! But he tried again, so I had to tell him, "Mr. Rogers, I am not interested in my personal comfort, only in the ability to do something!"

He said, "OK, go to NASA. It's OK." That's where our conversation ended.

So, I went. I got a private briefing all day at NASA on the engines and on the seals. The briefing on the seals was by Mr. Weeks. It was a continuation of my JPL briefing, with many more details, including the history of these matters: how the problem had been discovered very early, how there had been "burn-throughs," "erosion," "blow-bys" and what not, on flight after flight—how many there were, and how each flight readiness review had looked at the information and decided it was all right to fly.

At the end of this long report on the problem of the seals, there was a page with recommendations. This is how all information is communicated in NASA—by writing everything down behind little black circles, called "bullets."

When I looked at the recommendations, the thing that struck me was the contradiction between two of the bullets: The first one says, "The lack of a good secondary seal in the field joint is most critical. Ways to reduce the effects should be incorporated as soon as possible to reduce criticality." Then, further down the page, it says, "Analysis of existing data indicates that it is safe to continue flying with existing design . . . "—with some other conditions, such as using 200 lbs of pressure in the leak test. (By the way, we discovered later that the leak test itself was causing the holes in the putty and was part of the reason for the failure of the seals!)

I pointed out this contradiction and said, "What analysis?" It was some kind of computer model. A computer model that determines the degree to which a piece of rubber will burn in a complex situation like that—is something I don't believe in!

I also found out that the matters that were causing trouble were brought up only at the "flight readiness review," where they were deciding whether to fly or not. There are so many considerations in deciding whether to fly, yet they brought up these critical matters only under those circumstances. In between the flights, there was no discussion of the problem—how it's going along or whether there's some progress.

So, what was really happening was that NASA had developed an attitude: If the seals leaked a little and the flight was successful, it meant that the seal situation wasn't serious. Therefore, the seals could leak and it would be all right—it was no worse than the time before.

Such an attitude is, of course, extremely dangerous. One or two out of five seals leaked—and only some of the time—so it's obviously a probabilistic matter, a thing you have no control over, an uncertainty. And it's *not* obvious that the next time you fly, the uncertainty won't click over a little bit more, statistically, and the seal will fail. Try playing Russian roulette that way: You pull the trigger and it doesn't go off, so it must be OK to do it again, right?

The next morning, Sunday, Mr. Graham took me with his family to the National Air and Space Museum. There we saw a moving picture about NASA, and it was so well done that I almost cried when I saw all the people involved at every level, how enthusiastic everybody was and how eager they were to make things work. That made me even more determined to help straighten things out as quickly as possible and to talk to the shuttle assembly people, the engineers and everybody else low enough down.

Later that day, General Kutyna called me up on the telephone. "I was working on my carburetor, and I was thinking. You're a professor," he says. "What, sir, is the effect of cold on the rubber seals?"

I caught on immediately to what he was thinking of. The temperature was $29°$ when the shuttle flew, and the coldest previous launch was $53°$. I said, "You know as well as I do. It gets stiff and loses its resiliency." That gave me a clue. Of course, that's all he had to tell me, and it was a clue for which I got a lot of credit later. But it was his idea. The professor of physics always has to be told what to look for. You just use your knowledge to answer the questions.

That weekend, *The New York Times* put out an article about a man named Cook, who was in the budget department of NASA. Mr. Cook had written a letter to his superior a year earlier, saying that the engineers knew there was something wrong with the seals, that they might have to fix the problem, and it might be expensive. Mr. Cook was working out the budget and recommended that NASA prepare for the contingency that it would suddenly need a big load of money to fix this problem of the seals.

This gets into *The New York Times*, and so we have to have a special meeting. It's the press, you see; we have to match the press. So on Monday, everybody was called to a meeting anyway! But I remind you, we still hadn't had any meetings in which we did any work. At this emergency closed meeting, we got some interesting information: The NASA people who had been looking at the television pictures of the launch saw preliminary indications that there was smoke coming out of one of the joints just at liftoff.

More interesting still was a report by a man named MacDonald from the Morton Thiokol Company, who came to the meeting on his own. He said that the Thiokol Company engineers had *noticed* the low temperature, had been *worrying* about the seals, had *known* about the resilience not being there. Furthermore, they knew that when it is cold, the grease in the seals is very viscous so it can't move fast enough to close the gaps. The engineers

were very, very worried about it just before the flight and reported to the people at Marshall that they should not fly below $53°$
temperature, and that night it was $29°$. But the engineers were
told that that was an appalling decision, that they should think it
over again, and they were given some apparently logical reason.

We later learned that in the discussions inside Thiokol, the
engineers were *still* saying, "We shouldn't fly," but the managers
made a decision nevertheless to go ahead and fly, and then they
gave the usual, apparently logical reason, which was—never
mind, I couldn't ever understand it. It's hopeless.

At any rate, that morning I had asked the question about how
resilient the rubber is, and, as always, NASA was very cooperative
at giving me information. That afternoon I got a stack of papers,
the first page of which said, "Mr. Feynman of the commission
wants to know about the resiliency of the O-ring rubber at low
temperatures . . . "—and it's sent to the next subordinate. The
subordinate writes to another subordinate, "Mr. Feynman of the
Presidential commission wants to know . . . " and so on, down
the line. In the middle there's a paper with the answer, and then
there's a series of papers—the submission papers—which explain
that "this is in answer to your request at such-and-such a time."

So I get this stack of papers, just like a sandwich, and in the
middle the answer is given to the wrong question! The answer I
got was: when you squeeze the rubber for two hours at a certain
temperature and pressure, what happens when you let go—how
long it takes to creep back—over *hours*. And I was talking about
fractions of a second during launch when the gap in the field joint
is suddenly changing. So the information was of no use.

We were going to have a public meeting the next day. I was
already getting tired of these public meetings and briefings because they were so time consuming and of so little use. I thought,
"Now we're going to have an open meeting, and we're going to
say exactly the same thing that we did in the closed meeting." (It
was a good idea: Mr. Rogers wanted to keep the public informed,
so every time we discovered something, we would quickly have
an open meeting to bring out the new material.) But I thought,
"It's like an act: We have to hear the same things in the open
meetings as in the closed meetings, and we won't learn anything
new. And the information I got from NASA about the rubber is
useless."

Later I'm feeling lousy and I'm eating dinner; I look at the table, and there's a glass of ice water. I think, "Damn it, *I* can find out about that rubber *without* sending notes to NASA and getting back a stack of papers; all I've got to do is get a sample of the rubber, stick it in ice water and see how it responds when I squeeze it! That way, I can learn something *new* in a public meeting!"

I ask NASA for a piece of the rubber. It's impossible to get; they're very, very careful, and every piece of material is checked and counted and everything else, so you can't just go down to the stockroom and pick up a piece of rubber. But Mr. Graham remembered there were two pieces of the rubber in the field joint model NASA had shown us before and was going to use again in the open meeting. The two pieces of rubber were the real thing, about an inch and a half long each. We decided to meet in Mr. Graham's office the next morning before the meeting to see if I could take the model apart. (In the open meeting I would have to take the model apart quickly.)

The next morning I get up early. I come out of the hotel—it's snowing a little bit—and I'm dressed up in my suit because I'm going to the public meeting later. A taxi comes up, and I say to the driver, "I want to go to a hardware store."

He says, "A hardware store? There's no hardware stores here. The Capitol is just up the street—we're in downtown Washington!" Then he remembered where he had seen a hardware store once, some distance away, and we went there. It didn't open till 8:30—it was about 8:15—so I waited outside, in my suit coat and tie, a costume I had assumed since I came to Washington in order to move among the natives without being too conspicuous.

The suit coats that the natives wear inside their buildings (which are well heated) are sufficient for walking from one building to another, or from a building to a taxi, if the distances are too great. (All the taxis are heated.) But I observed that the natives seem to have a strange fear of the cold: On top of their suit coats they put on overcoats if they wish to step outside. I hadn't bought an overcoat yet, so I was still rather conspicuous standing outside the hardware store in the snow.

When the hardware store opened, I bought myself some screwdrivers, pliers, clamps and so on, because I wasn't sure exactly what I would need.

When I got to NASA I began thinking the clamps were too big to put into a glass. So to get some small clamps I went to the medical department of NASA, where I had gone several times before (my cardiologist was trying to take care of me by telephone). I went up to Graham's office. He was very cooperative, as always, and we saw that I could open the model very easily with just a pair of pliers. So there was the rubber, right in my hand, and although I knew it would be more dramatic and honest to do the experiment directly in the meeting, I cheated—I couldn't resist. I tried it. And, after all, it would be quite a flop if it didn't work! So, following the example of having a closed meeting before an open meeting I must tell you I discovered it worked before I did it in the open meeting.

I kept wanting to do my experiment all during the meeting, but General Kutyna, who was sitting next to me, gave me advice. He had given me advice before. At the first public meeting he had leaned over and said, "Copilot to pilot: Comb your hair." So now he was saying, "Copilot to pilot: Not now!"

So when he told me, "Now!" I did it, and everything went all right. As you probably know, I demonstrated that the rubber had no resilience whatever when you squeezed it at that temperature, and that it was very likely a partial cause of the accident. We all agreed later that that, in fact, was true.

On Wednesday, 12 February, we had no meeting, so I wrote a letter home. I told my wife she was right, that in certain ways I *was* unique. One of the ways I was unique was that I was not connected to any organization—I had no weakness from that point of view. I was, of course, connected with Caltech, but that's not a weakness! For example, General Kutyna was in the Air Force, so he couldn't say everything exactly the way he wanted, because he might get in trouble with the Air Force. Sally Ride still had a job at NASA. Everyone on the commission had some kind of connection and therefore some kind of weakness, but I was apparently invincible.

But General Kutyna warned me that when they fly airplanes, they have a rule: Check six. Most airplanes are shot down this way: A guy is flying along, looking in all directions, and feeling very safe. An airplane flies up behind him (at "six o'clock"; "twelve o'clock" is directly in front), and he gets hit. So you always have to check six o'clock. So I began to write, "Check six!" on every note paper I had and developed a kind of paranoia.

For example, I have a cousin who previously had been with the Associated Press as White House correspondent and is now with CNN; I also have a nephew who works for *The Washington Post*. When I had some time I would visit with them—eating dinner, and so on. It was very pleasant, but we made sure we never said a word about anything I was doing, because I didn't want to be responsible for any leaks. I told Mr. Rogers that I had these associations with the press. He smiled and said, "It's perfectly all right. I used to work for so-and-so"—he had some connection with the press too. He just laughed; there was no problem. But my paranoia had developed to such a point that I thought, "That was too easy; he's going to get me that way!" So I stopped seeing my cousin. That was stupid: There were no problems; it was just my state of mind.

I did, however, keep talking to the press—openly, always giving my name. I didn't want any hocus-pocus about "unidentified sources," or anything. My cousin had taught me that the press is not something to be afraid of, and it turns out to be true. I found that out several times. The first time was when *The New York Times* put out an article after I did the ice water experiment; during the public meeting I had no time to explain what its meaning and importance were, but *they* had it all explained perfectly.

Another time, NBC interviewed me—they caught me in the lobby of my hotel. They interviewed me for 15 to 20 minutes—the lady reporter was very short and very nice—and I talked in my usual, careful, professional way, with all the caveats and so forths and so ons. I saw the interview later on the "Nightly News": I was on for about two seconds—I say something, and BOOM!—it's over. But it was good: The report carried the line of what I said, and the reporter put the context around it, saying things like, "The professor went on to say that this was only the result of a mathematical model and might be uncertain"—stuff like that. It was excellent. It was very short, carefully put together and excellent—except for one thing: Because I'm not experienced, I didn't look into the camera when I spoke. Instead, it looked like I was talking to my dog.

Well, finally, on Thursday, we get to Kennedy. The main briefing turned out to be the way I thought it would be—we didn't get any useful information just looking around at the "gee-whiz"place. But before that, we had two meetings in which we got a lot of information. We got a detailed look at the pictures of the smoke, which made it very apparent that the leak of gases through the seal had started immediately after ignition, then somehow plugged itself up temporarily, and finally ended up with a flame coming through. We also got all the details on the Thioko.-Marshall discussions, in which the engineers never changed their minds; only the manager did, under pressure from Marshall.

After two days at Kennedy, we were supposed to return to Washington. I thought, "Now, at last, here I am. Now I've got a chance to talk to everybody."

I told Mr. Rogers I wanted to stay at Kennedy, and he said, "I'd prefer that you didn't stay down here, but of course you can do whatever you want."

I said, "Well, OK, then, I'll stay."

So I stayed at Kennedy a few more days. I ran around and found about more about the pictures from the photograph guys; I found out about the ice that had been on the launch pad from the ice crew. They told me they had gotten some funny numbers for the temperature on the morning of the launch, and we discussed what was wrong. We called up the people who made the instrument, and tried to find out how the instrument was built so we could understand the errors, but they suddenly clammed up, obviously afraid that they were going to be blamed for the shuttle disaster.

I explained to the manufacturer that the instruments were not used in accordance with their manual (they had been used too soon after being taken out of the box), and we wanted to know what the effect of that misuse would be on the apparent temperature readings, and so forth. I finally got them to explain it all. They said our errors were reproducible. So we set up an experiment in which we reproduced the circumstances, and we corrected the temperature readings. I'm only trying to say I was working hard.

Another thing came up while I was running around down there at Kennedy. I had predicted that Mr. Rogers was going to try to fix me by overloading me—by giving me a lot of stuff to do. Sure enough, it happened; the commission staff in Washington kept sending me things to do. But as the instructions came in, I had done them already—they didn't realize how fast I am at getting information and understanding it and going on to the next thing.

The only thing they sent me that I didn't do had to do with a certain memo whose existence they had discovered. During the assembly of the solid rocket boosters, someone had written cavalierly, "Let's go for it!" The staff didn't like the attitude on the part of the workers, and they wanted to find that piece of paper. By that time I knew how much paper there was in NASA so I was sure it was a trick to make me get lost and to do nothing. So I did nothing about it.

I talked to Mr, Lamberth, who was in charge of the assembly of the SRBs. He told me about the problems he had with the workmen. They had a little accident earlier, and he had to discipline them about it, and then he told me about another incident: The SRBs become a little bit out of round after each use. When the workers were trying to make the rocket round again with the rounding machine—a rod with a hydraulic press on one end and a nut on the other—they were only supposed to go up to 1250 lbs, according to the manual. But they couldn't get it squashed enough that way, so they took a wrench and tightened the nut on the other end of the rod to squeeze it some more. That made the rocket round, all right, but one of the workmen noticed that the pressure had gone up to 1350 lbs that way. Well, a gauge measures the force applied to a rod from *either* end, so tightening the nut increases the pressure past 1250 lbs, of course! So Mr. Lamberth admonished the workers to follow the manual. He said the workers weren't like they used to be, and he was very disturbed.

So I go down and talk to the workers. First of all, I'm surprised to find that the foreman doesn't know anything about this admonishment. He knew about the 1350 lbs, but he didn't know he had been admonished. He said, "No, we weren't admonished; we were following the procedures in the manual." Sure enough, the manual said to tighten the nut after the pressure reaches 1250 lbs—it said so in black and white! It didn't say that tightening the nut would increase the pressure; the people who wrote the manual probably weren't quite aware of that. So the workmen had, in fact, followed the manual perfectly. (I later found out that as a result, the manual was revised to allow for higher pressure, and that only the hydraulic jack was to be used to increase the pressure. The step about tightening the nut was eliminated.)

So Mr. Lamberth really didn't know what happened underneath. He said he had admonished the workmen, but he never talked to them directly. So he had the idea that his workmen were no longer like they used to be, but I tell you, they really were. They had a lot of information but no way to communicate it. The workmen knew a lot. They had noticed all kinds of problems and had all kinds of ideas on how to fix them, but no one had paid much attention to them. The reason was: Any observations had to be reported in writing, and a lot of these guys didn't know how to write good memos. But they had good knowledge, they worked very hard and they were very enthusiastic.

While I was doing my work down at Kennedy, Mr. Rogers was in Washington appearing before a Congressional committee. (Congress was considering whether to set up its own investigation of the accident.) Senator Hollings said, "So who have ya got, there, on your commission? Ya got a couple of astronauts, a Nobel prizewinner, a general, some businessman and a couple of lawyers. What you really need is gumshoes, who will be right down there at Kennedy, eating lunch with the very guys who do the work on the shuttle."

And Mr. Rogers was able to reply, "You'll be interested to know, Senator, that the Nobel prizewinner is down there at Kennedy, right now, doing exactly that!" (Although Mr. Rogers couldn't have known it, I was actually eating lunch with some of the engineeers at exactly that time.) So Mr. Rogers gradually realized I wasn't quite so useless. We got to respect each other very much—I think he ultimately respected me, and I certainly do respect him for his abilities.

I went back to Washington, and I got into more and more difficulties. The next meeting we had was a public meeting, and I was questioning Mr. Lund of the Thiokol Company, who had changed his mind about launching the shuttle. Somebody at Marshall had told him to put on his "management hat" instead of his "engineering hat," and so he changed his opinion. I was asking him, "Don't you understand the principles of probability?" when suddenly I had this feeling of the Inquisition.

Mr. Rogers had pointed out to us that we ought to be careful with these people, whose careers depended on us. He said, "We have all the advantages. We're sitting up here, they're sitting down there; they have to answer our questions, we don't have to answer their questions. It isn't fair." Suddenly all this came back to me and I felt terrible. I couldn't do it the next day, so I went back to California, just for a day or two, to rest up.

While I was in Pasadena, I went over to JPL and discussed the enhancement of the pictures with Jerry Solomon and Meemong Lee; they were studying the flame that had appeared on the side of the SRB just before the main fuel tank exploded. I had just been in Washington, hearing the NASA managers talk through a fog. What a difference—just like with the photograph guys and the ice crew at Kennedy, everything was so direct and simple at Caltech and JPL. What a difference!

We finally divided into working groups, and I went to Marshall with General Kutyna's group. The first thing that happened there was, a range safety officer by the name of Ulian came to tell us about a discussion he had had with NASA higher-ups about safety. Mr. Ulian had to decide whether to put explosive charges on the side so ground control could destroy the shuttle in case it was falling onto a city. The big cheeses at NASA said, "Don't put any explosives on, because the shuttle is so safe. It'll never fall onto a city."

Mr. Ulian tried to argue that there *was* danger. One out of every 25 rockets had failed previously, so Mr. Ulian estimated the probability of danger to be about 1 in 100—enough to justify the explosive charges. But the higher-ups at NASA said that the probability of failure was 1 in 100,000." That means if you flew the shuttle *every day*, the average time before your first accident would be 300 years—every day, one flight, for 300 years—which is obviously crazy! Mr. Ulian also told us about the problems he had with the big cheeses—how they didn't come to the meetings sometimes and all kinds of other details.

Then I thought of this question: By now we had found out that the flight failed because one of the seals had broken, and the higher-ups had told us they didn't know anything about the seals problem—even though I was able to find out about it right away at JPL, before I even went to Washington. We saw that NASA had no system for fixing the problem, even though engineers were writing letters like, "HELP!" and "This is a RED ALERT!" Nothing was happening. My question was: Does this lack of communication between engineers and management also exist in other places? I thought, "I oughta find out whether this is a characteristic of the whole system, or whether it's true just for Morton Thiokol, and we happened to find out about it because the O-rings busted." So I told the people at Marshall I wanted to find out about the engines. I wanted to talk to a couple of engineers without any managers around.

"Yes, sir, we'll fix it up. How about tomorrow morning at 9:00?"

The next day I come in, and there's engineers, all right, but there's also managers, and a great, big book: *Presentation Made on February Such-and-Such to Commissioner Richard P. Feynman*—all prepared during the night.

"Geez! It's so much work!" I said.

"No, it's not so much work; we just put the regular papers in that we use all the time."

The engine is extremely complex and hard to understand, and the engineers were explaining to me how it worked, showing slide after slide. I asked my usual dumb-sounding questions.

After a while, Mr. Lovingood, a middle manager there, said, "Mr. Feynman, we've been going for two hours now. There are 123 pages, and we've only covered 20."

"It's all right, don't worry," I said. "I'm confident that it'll go faster as we go along, but I want my questions answered at the beginning. Otherwise, I can't understand it."

Suddenly I got an idea. I said, "All right, I'll tell you what. In order to save time, the main question I want to know is this: Is there the same understanding, or difference of understanding, between the engineers and the management associated with the engines as we have discovered associated with the solid rocket boosters?"

Mr. Lovingood says, "No, of course not. Although I'm now a manager, I was trained as an engineer."

I gave each person a piece of paper. I said, "Now, each of you please write down what you think the probability of failure for a flight is, due to a failure in the engines."

I got four answers—three from the engineers and one from Mr. Lovingood, the manager. The answers from the engineers all said, in one form or another (the usual way engineers write—"reliability limit," or "confidence sub so-on"), almost exactly the same thing: 1 in about 200. Mr. Lovingood's answer said, "Cannot quantify. Reliability is determined by studies of this, checks on that, experience here"—blah, blah, blah, blah, blah.

"Well," I said, "I've got four answers. One of them weaseled." I turned to Mr. Lovingood and said, "I think you weaseled."

He says, "I don't think I weaseled."

"Well, look," I said. "You didn't tell me *what* your confidence was; you told me *how* you determined it. What I want to know is: After you determined it, what *was* it?"

He says, "100 percent." The engineers' jaws drop. My jaw drops. I look at him, everybody looks at him—and he says, "Uh . . . uh, minus epsilon?"

"OK. Now the only problem left is, what is epsilon?"

He says, "1 in 100,000." So I showed Mr. Lovingood the other answers and said, "I see there *is* a difference between engineers and management in their information and knowledge here, just as there was in the case of the rocket, but let me not bother you about it; let's continue with the engine."

So they continued telling me about the engine, and soon I understood how it worked. Then they told me about all the problems they had had with it—blades cracking, and all kinds of other difficulties. And I discovered the same game, just as in the case of the solid rocket boosters, of reducing criteria and accepting more and more errors that weren't designed into the device.

Later I also checked the avionics, the software NASA uses on its computers for controlling the shuttle from launch to landing, to find out if a similar situation existed there. But in this case, on the contrary, everything was very good; the engineers and the managers communicated well with each other, and they were all very careful not to change their criteria of acceptance during flight reviews. I found the avionics completely satisfactory.

I wrote up what I found out about these things into a special report, hoping that the other members would see it for discussion. I sent it to Al Keel, the executive officer whom Mr. Rogers had selected to coordinate everything on the commission. He told me on the telephone that he had received it and that he would show it to everybody.

By this time we were beginning to write up our part of the main report about the accident. General Kutyna had set up a whole system at Marshall for doing so. It lasted about two days before we got a message from Mr. Rogers: "Come back to Washington. You shouldn't do the writing down there." So we went back to Washington, and Mr. Graham lent me an office and a secretary who was very, very good. I helped our group write up its part of the main report—with a lot of input from Mr. Keel.

All this time I had expected that we would be meeting in Washington to discuss what we had found out so far, to think it out together and look at it from different perspectives—in addition to the astronauts there were lawyers and industrialists, there were scientists and engineers, and so on—and to discuss with each other where to go next. But in our meetings, all we ever did was what they called "wordsmithing"—correcting punctuation, refining phrases and so on. We never had a real discussion of ideas!

Besides the wordsmithing, we discussed the typography and the color of the cover. At each meeting we were asked to vote, so I thought it would be efficient to vote for the same color we had decided on in the meeting before—but it turned out I was always in the minority! We finally chose red. It came out blue.

At any rate, after one of the meetings I was talking to Sally Ride about my experiences investigating the engines and the avionics, and I noticed that she didn't seem to know about the special report I had written—the one Mr. Keel told me he would show to everybody. So I said to Mr. Keel, "Sally hasn't seen my report."

He says to his secretary, "Oh, make a copy of Mr. Feynman's report and give it to Ms. Ride."

Then I discovered Mr. Acheson hadn't seen it.

"Make a copy and give it to Mr. Acheson."

I finally caught on, so I said, "Mr. Keel, I don't think anybody has seen my report."

So he said to his secretary, "Make a copy for all the commissioners and give it to them."

Then I said, "I thought you told me you showed it to everybody."

"I meant I showed it to the entire staff."

Needless to say, when I asked the members of the staff about it, none of them had seen it either.

When the commissioners read my report, all of them thought there was a lot of good stuff in it, and it ought to be in the commission report somewhere. But we couldn't discuss it, because all we were doing was this wordsmithing stuff on what was already written—not adding anything new. We were working on the summary report for the President—I'll call it the main report—which was relatively brief. Later, as backup data and other information, we were going to put out a series of appendices. So, I thought, there are two possibilities for my report. It could be in the main report—but it would have to be rewritten in that case, because the style of the main report was different—or it could be put out later as an appendix.

Although some of the members felt strongly that it ought to go in the main report, I thought I'd compromise, and let it go in as an appendix. But in order to get my report in as an appendix, it had to be put into the document system computer, which was quite elaborate and very good, but different from the computer system I had written my report on at home. They had an optical scanner for transferring it, so I asked them to do that, and they said, "Of course."

I'd go away for a while, and when I'd come back, it would be lost. But I kept pushing on it, watching it, nursing it along, and I finally got it through to the point where it was, at last, in the hands of a real editor, a capable man by the name of Hansen, who changed all my *which*es to *that*s and *that*s to *which*es.

Mr. Hansen fixed up my report without changing the sense of it. Then Mr. Keel fixed it up so it could go in as an appendix: He put all kinds of big circles around whole sections, with Xs through them; there were all kinds of thoughts left out. He explained to me that my report was repetitious with the main report, and I argued that it's much easier to read something that's all together, and because it was going to be an appendix, repetition didn't matter.

Finally, the commission had its last meeting. It was about the recommendations we would make to the President. We made nine recommendations. The next day, I'm standing around in Mr. Rogers's office when he says, "I thought we would add a tenth recommendation: 'The commission strongly recommends that NASA continue to receive the support of the Administration and the nation . . . '" In our four months of work as a commission, we had never discussed that issue. It wasn't in our directive from the President. We were only to look at the accident, find out what caused it and make recommendations to avoid such accidents in the future.

So I thought this tenth recommendation wasn't appropriate and said so. We argued back and forth a little bit, but then I had to catch a plane to New York, where I was going for the weekend. While I was in the airplane, I thought about it some more, and the more I thought about it, the more I thought what a mistake it looks like—just like one of the NASA reports, like the one I had seen back at the beginning, with the contradictory bullets: There's all these troubles, but in the end we recommend to keep on flying!

I knew I didn't like it. Furthermore, we hadn't discussed it at a meeting! It was just Mr. Rogers's idea. I didn't want to call up Mr. Rogers and argue with him on the telephone, so I quietly and thoughtfully wrote out a letter to him, carefully explaining why I didn't like the tenth recommendation. To make sure it got there right away, I dictated my letter over the telephone to Mr. Rogers's secretary, who typed it up and handed it to him right in his office!

When I came back from New York, Mr. Rogers told me that he had read my letter. He said he agreed with it, but that I was outvoted.

I said, "How was I outvoted, when there was no meeting?" I thought my ideas about this were worth discussing with the other commissioners, and I wanted to know what they thought about my arguments.

"I know, but I called each one of them up," he said, "and they've all agreed. They've all voted for it."

So I said, "Well, I'd like a copy of this recommendation," and I went off to make a copy of it. When I came back, Mr. Keel said he forgot that they hadn't talked to Mr. Hotz—Mr. Hotz was there, you see, so I could ask *him* right away. They forgot that they hadn't talked to Mr. Hotz. I went to lunch with Mr. Acheson and Mr. Hotz, and it seemed like Mr. Hotz agreed with me. When we went back to Mr. Rogers's office, Mr. Acheson explained to me, "It's only 'motherhood and apple pie.' If this were a commission for the National Academy of Sciences, your objections would be proper. But since this is a Presidential commission, we should say something for the President."

"I don't understand the difference," I said. (Being naive at the right time is often a good idea.) "I just don't understand. Why can't I be careful and scientific when I'm writing a report to the President?" (Being naive doesn't always work: My argument had no effect.)

I was very concerned by all this, and I came home for a while, very disturbed. I then got the idea—which I hadn't had before—to call up some of the other commissioners. I'll call them A, B and C.

I call A. He says, "What tenth recommendation?"

I call B. He says, "Tenth recommendation? What are you talking about?"

I call C. He says, "Don't you remember, you dope? I was in the office when Rogers first told us, and I don't see anything wrong with it."

Although some of the commissioners agreed with the tenth recommendation, I still thought we should have discussed it in a meeting. I had also been railroaded into modifying my report, even though it was going to appear only as an appendix. I talked to my sister, who used to work in Washington.

She said, "Well, if they do that to your report, what happens to all the work you did on the commission? Your contribution wouldn't be seen. It would appear as if you didn't do anything."

I said, "Aha!" and I sent a telegram to Mr. Rogers:

PLEASE TAKE MY SIGNATURE OFF THE FRONT PAGE OF THE REPORT UNLESS TWO THINGS OCCUR: 1) THERE IS NO TENTH RECOMMENDATION, AND 2) MY REPORT APPEARS AS AN APPENDIX WITHOUT MODIFICATION FROM VERSION # 23 of MR. HANSEN.

I knew by this time I had to define everything carefully! (By the way, *everything* had 23 versions. It has been noted that computers, which are supposed to increase the speed at which we do things, have not increased the speed at which we write reports. We used to make only three versions—because they're so hard to type—and now we make 23 versions!)

The result of this telegram was that Mr. Rogers and Mr. Keel tried to compromise. They asked General Kutyna to be the intermediary, because they knew he was a friend of mine. What a *good* friend of mine he was, they didn't know.

The general calls me up, and right away he says, "Hello, professor! Let me first tell you, I'm with you. But I've been given the job of convincing you to change your mind, and I have to give you all the arguments."

"Fear not!" I said. "I'm not gonna change my mind. Just give me the arguments, and fear not."

So he gave me all the arguments, none of which had any effect. The arguments were all kinds of crazy things. For example, "If you don't accept the tenth recommendation, they're not going to accept the compromise they already made about putting your report in as an appendix." I didn't worry about that one, because I didn't have to sign the main report, and I could always put out my report by myself.

Another argument was that they noticed I was always talking to the press and they would claim I was doing this as a publicity stunt to sell more copies of my book. That one made me smile, because I could imagine the laughter it would produce from my friends at home. I knew that nobody I cared about would believe it.

But finally, I did compromise. I said, "Instead of making it a recommendation, just make it a concluding thought and change the wording from 'strongly recommends' to simply 'urges.'"

They accepted that.

A little bit later, Mr. Keel calls me up: "Can we say *'strongly urges'*?"

I said, "No. Just 'urges.'"

So I put my name on the main report, my report got in as an appendix, and everything was all right. We gave our report to the President on a Thursday in a ceremony at the White House in the Rose Garden. The report was not to be publicized until Monday, so the President could study it.

During those three days the newspaper reporters were working like demons. They knew the report was finished, and they were trying to scoop each other to find out what was in it. They kept calling me up because I had been so cooperative before. I told my secretary to say that I had no comment on anything; I would answer all their questions on Tuesday at my news conference.

Well, I didn't know it, but someone had leaked that this argument had gone on. The only man who knew about it, I think, was Mr. Hotz. He may have thought it would help me in pushing my point, but for whatever reason, it leaked. Some paper in Miami started it, and soon the story was running all over about this argument between me and Mr. Rogers. So when the reporters called me up, they'd get the message, "He has nothing to say; he'll answer all your questions at his press conference on Tuesday."

That sounded very suspicious, so my press conference turned out to be very popular. That's what most of the questions at the news conference were about. So I found myself repeating that I don't have any problem with Mr. Rogers.

My reaction is that I like him, and that he's a genuinely fine fellow. But I reserve in my head the possibility—not as a suspicion, but as an unknown—that I like him because he's such a good politician that he knew how to make me like him. I prefer to assume he is the way he appears. But as a scientist, I'm not sure that my evidence is complete. And I was in Washington long enough to know that I can't tell.

Finally, I would like to say something about the general deterioration of NASA—and the fact that there was no information coming up from the engineers to the management. Just the other day I was reading a book by Harvey Brooks in which he talked about innovation. He explained that innovation doesn't have to be the direct invention of a machine; an innovation could be the was things are made, such as the Ford mass production line or, as in another of his examples, the management system developed at NASA for the Apollo program, which involved the cooperation of so many contractors and subcontractors. The system they evolved was an innovation, a great development. This was more than 20 years ago. But in the meantime, something happened that happens to many human innovations—it deteriorated. The question is: How and why? I don't know.

I invented a theory, which I have discussed with a considerable number of people, and many people have explained to me why my theory is wrong. But I don't remember their explanations as to why it's wrong—you never can, because that's the way you're built! I am a weak human, too, so I cannot resist telling you what I think is the problem.

When NASA was trying to go to the Moon, it was a goal that everyone was eager to achieve. Everybody was cooperating, much like the efforts to build the first atomic bomb at Los Alamos. There was no problem between the management and the other people, because they were all trying to do the same thing. But then, after going to the Moon, NASA had all these people together, all these institutions and so on. You don't want to fire people and send them out in the street when you're done. So the problem is what to do.

You have to convince Congress that there exists a project this organization can do. In order to do so, it is necessary (at least it was *apparently* necessary in this case) to exaggerate—to exaggerate how economical the shuttle was going to be, to exaggerate the big scientific facts that would be discovered. (In every newspaper article about the shuttle there was a statement about the useful zero-gravity experiments—such as making pharmaceuticals, new alloys and so on—on board, but I've never seen in any science article any results of anything that have ever come out of any of those science experiments which were so *important!*) So NASA exaggerated how little the shuttle would cost, they exaggerated how often it could fly, to such a pitch that it was *obviously incorrect*—obvious enough that all kinds of organizations were writing reports, trying to get the Congress to wake up to the fact that NASA's claims weren't true.

I believe that what happened was—remember, this is only a theory, because I tell you, people don't agree—that although the engineers down in the works knew NASA's claims were impossible, and the guys at the top knew that somehow they had exaggerated, the guys at the top didn't want to *hear* that they had exaggerated. They didn't want to hear about the difficulties of the engineers—the fact that the shuttle can't fly so often, the fact that it might not work and so on. It's better if they *don't* hear it, so they can be much more "honest" when they're trying to get Congress to OK their projects.

So my theory is that the loss of common interest—between the engineers and scientists on the one hand and management on the other—is the cause of the deterioration in cooperation, which, as you've seen, produced a calamity.

TWO MINUTES[3]

No one who was watching the news on Jan. 28, 1986 will ever forget the sight. Seventy-three seconds into its flight, the *Challenger* spacecraft blew up, killing its seven-person crew.

The cause of the tragedy soon became clear: An O-ring seal, still cold because of sub-freezing temperatures the night before lift-off, had failed to do its job, allowing flames from inside the boosters to burn through the field joint and ignite the giant external tank.

[3]Article by Tom Bancroft. *Financial World.* 158:28+. Jn 27 '89. Copyright © 1989 by Financial World Partners. Reprinted with permission.

Challenger's solid rocket boosters were made at a Morton Thiokol factory about 100 miles north of Salt Lake City at the foot of the great Wasatch Mountains. Some of the employees of Thiokol's Wasatch plant were not all that surprised by the tragedy. Certainly there had been plenty of warnings.

As far back as Oct. 21, 1977, an engineer at NASA's Marshall Space Flight Center in Huntsville, Ala., handed in his evaluation of the O-ring joint designed by Morton Thiokol. His conclusion: The design was "unacceptable." But NASA paid little attention to that warning.

The engineer, Leon Ray, then took his findings to two companies that manufacture O-ring seals, Parker Seal and Precision Rubber Products. Their initial reaction, he reported, was that the "O-ring was being asked to perform beyond its intended design, and that a different type of seal should be considered." All NASA did, in 1979, was to begin investigating second sources for the solid rocket booster. But because of "budgetary priorities," funding for a competition was put on hold.

As successive shuttle launches proceeded, serious problems with the O-ring seals occurred on at least nine flights before 51-L, the *Challenger*. The first evidence of problems with the seals appeared during the *Columbia* flight in November 1981, but Thiokol ignored it. In its report to the President, the Rogers Commission, specially appointed to investigate the *Challenger* tragedy and recommend any changes that might be necessary in the space program, wrote, "Morton Thiokol, the contractor, did not accept the implications of tests early in the program that the design had a serious and unanticipated flaw." The red flag finally went up in January 1985 with flight 51-C, the *Discovery*. Roger Boisjoly was one of the first engineers at Morton Thiokol to notice real damage to the O-ring seals, before and after the 51-C flight. It was also clear by this point that low launch temperatures seriously aggravated the O-ring problem. Still, neither NASA nor Thiokol did anything more than identify the problem as critical.

In a paper he wrote to the American Society of Mechanical Engineers, Boisjoly says, "January 1985 was the period of gross escalation of the joint seal problems. After that there were events that should have happened that would have stopped the launches, or changed the launch commit criteria to prevent launching below 53 degrees."

That summer, Boisjoly tried to form a seal-erosion task force to correct the problem, but ran into roadblocks. In his paper he wrote, "The team was frustrated from the start due to lack of management support to provide manpower and material resources."

In October 1985, Robert Ebeling, manager of the booster ignition systems, and a member of the seal task force, wrote a memo to Thiokol management desperately trying to get attention. "HELP!" he wrote. "The seal task force is constantly being delayed by every possible means." Another letter, written by a Thiokol engineer named Scott Stein, said virtually the same thing. Of the next 10 flights to follow 51-C, there were eight instances of hot gases penetrating the field joint or nozzle joint seals. On flight 51-L the morning of Jan. 28, 1986, a seal on the aft field joint of the right solid rocket booster completely gave way, allowing a flame to penetrate the external tank. The *Challenger* exploded.

The record raises serious ethical questions: Knowing of the repeated problems with the O-rings, should Morton Thiokol's top management have taken remedial action? And having had good reason to believe the O-rings tended to malfunction at low temperatures, should they have given their approval to launch?

The top management of Morton Thiokol declined to be interviewed for this article.

The Rogers Commission quickly reconstructed what happened immediately prior to launch. There had been enormous pressure to get the spacecraft airborne. Flight 51-L was already a year behind schedule. It had been postponed yet again because of high winds on Jan. 27. The night before the launch, Morton Thiokol engineers Allan McDonald, Arnie Thompson, Roger Boisjoly, Bob Ebeling and others, after learning of freezing-cold temperatures predicted for Cape Kennedy that night, pleaded with Thiokol and NASA management not to launch. An O-ring, which is rubber, loses its resiliency when it gets cold, and therefore its ability to seal the joint properly.

Originally, according to the commission's report, Thiokol managers agreed, telling NASA to scrub the launch. But a NASA manager named Larry Mulloy argued that the evidence presented by Thiokol's engineers was inconclusive. Then, Jerry Mason, senior vice president of Thiokol at the time, asked his head of engineering, Bob Lund, to "take off his engineer's cap, and put on his management cap," according to the report. Thirty minutes later, despite vehement protests from their top solid rocket booster engineers, Thiokol gave its recommendation to launch.

A defensible decision? The commission called for expert technical assistance. They ordered that veteran engineers with aerospace experience be allowed to scrutinize contractor operations in depth and report back to NASA all conceivable hazards related to the shuttle transportation system. On Sept. 22, 1988, the *Wall Street Journal* reported that Morton Thiokol had admitted to reviewing 1,027 possible problems raised by the Rogers Commission engineers. Some of the engineers themselves suggest that over 2,000 hazard reports were filed.

Morton Thiokol's Utah facility alone had to hire roughly 60 engineers to look over the solid rocket boosters.

Right from the start, the outside engineering team ran into strong opposition. Before long, several of the Rogers Commission engineers began to complain that their reports were being suppressed by Thiokol. Some may never have reached NASA, one engineer warns. From the minute they set foot in the Wasatch plant in northern Utah, their hazard reports were censored and they were treated as outcasts. In his paper to the Society of Mechanical Engineers, Boisjoly writes that Morton Thiokol's "management style would not let anything compete or interfere with the production and shipping of its boosters. The result was a program which gave the appearance of being controlled, while actually collapsing from within."

Rogers Commission engineer Steve Agee said, "We were forbidden to speak up at meetings." Kent McKinnon, a systems safety supervisor, in a letter to Joyce McDevitt, head of NASA's Safety Risk Assessment Ad Hoc Committee, said he was told to keep silent when NASA representatives were at the plant. He wrote that if he had mentioned that there were problems with management, he had better "start looking for a job, because management would get rid of me."

According to Agee, few of the hazard reports written by the contractors were being sent to NASA. "It didn't matter if the problem was simple or grandiose, they didn't want to tell NASA there were problems of any shape or kind."

"If a hazard couldn't be resolved, my directions were not to even submit it," said Ron Clary, another engineer commissioned to work at Thiokol in March 1987. When Clary protested, he was ordered to cite examples of how safety information was being suppressed. "That would have taken me the rest of my life," he scoffs.

Just getting the relevant information on which to base their documentation of hazards was a huge problem. "It was withheld and tended to change practically every hour," Clary says.

A February 1987 interoffice memo writeen by P.R. Dykstra, vice president of strategic planning, suggests that Thiokol management was well aware of the cover-up. Paraphrasing Pogo he writes, "We have met the enemy, and they is us." He goes on to chastise, "superficial, relatively painless solutions to complex problems are not likely to resolve the problem." In an interview with FW, Dykstra said the memo was just a joke.

Eventually, in frustration Agee went to the FBI, turning over numerous documents and once, at their suggestion, even wore a wiretap to a meeting. The FBI is currently investigating the allegations of Agee and his colleague, Tony Laine, about Morton Thiokol. In a resignation letter dated Sept. 19, 1986, Cindy Ferrari, an employee in structures design, wrote that the company as a whole has "no real concern for their employees, and their response to complaints is basically, 'If you don't like it, quit.' Employees are frustrated because their recommendations are often ignored. Life is too short to spend it working for a company which does not know what they are doing."

Laine, who has a lengthy career in aerospace and electromagnetic engineering, alleges he was fired by Thiokol for trying to examine the electromagnetic characteristics of the boosters. In a lawsuit for wrongful discharge against the company, he claims he was told not to work on electromagnetics, and only to "cut and paste" old safety reports. Thiokol claims Laine was not asked to study the electromagnetics of the booster.

Agee, before he left, had written over 200 hazard reports, many of which he marked critical. Ross Bowman, Thiokol's director of safety and quality assurance, is on record as stating Agee's work was reviewed and rejected by a Thiokol safety professional. A number of Agee's reports focused on improper grounding on railroad cars that took the booster segments down to the Kennedy Space Center, and during other segments of the booster process.

"If it had been a normal Cessna aircraft, you would have run a wire from A to B to C," Agee explains. "When there are fixes available that cost $1.98, like the grounding in certain instances, and they don't get fixed, something is wrong."

Ron Clary also remarked on the grounding, calling it a pervasive concern. "The whole time I worked there, I never saw a good ground on anything," he says.

Another problem that Agee saw during his tenure was a reckless use of "waivers and deviations." These are normal processes used in any contract situation when plans change due to unforeseen obstacles. But during his time at the Wasatch plant, he claims, "NASA bought over a thousand waivers and deviations. They needed waivers and deviations like a drug addict needs his drugs."

Agee, Boisjoly, Laine and others say they were harassed by the company in every way imaginable. "I sat next to a guy for about two months who was using a cardboard box as a desk," remarks Agee.

Clary says that when he first started working there, static-proof suits designed to permit the engineers to work in certain highly flammable areas of the plant were not given to the engineers—effectively preventing them from working in those areas. Eventually, he says, the contracted engineers began stealing them after hours so they could do their job.

Why didn't all the other Rogers Commission engineers complain? "It is a given in the contracting business to keep your head down," says Clary. "I had to get out of the field." Clary is now working in Maryland for a construction company.

Clary feels that one of the reasons Thiokol's Utah plant is profitable is because it pays such low wages. But as a result, he argues, it has difficulty attracting the most talented people. In a July 1988 Ad Hoc Committee status report, the committee noted, "There is still some concern that the depth of experienced, skilled and competent people is too shallow."

In just the past five years, there have been six incidents of buildings at the Wasatch plant either blowing up or catching fire. Even in the solid fuel business, where a spark can set off 100,000 pounds of explosives in a second, such a high accident rate is unusual and suggests that major operational changes may have been necessary.

A General Accounting Office report, released in 1986, cites Thiokol's correction of safety problems as slow and incomplete. The report goes on to say that "Thiokol's implementation of actions to correct problems . . . is dismal."

In March 1986, a building at the Utah facility that was used to process fuel for Trident missiles blew up. The *Wall Street Journal* reported that in July 1987 the Air Force began withholding payments on its MX missile contract with Thiokol because of what it called poor workmanship.

That December, five workers were killed when the solid fuel in an MX missile "casting pit" ignited beneath them. Thiokol was fined $31,700 by the Utah Department of Occupational Safety and Health, and subsequently settled out of court for $14,700.

A UPI article reported that in May 1988 an Air Force plant representative at Thiokol determined that the company was not meeting safety and cleanliness standards, resulting in $4 million of withheld payments once again.

Did Morton Thiokol's inattention to the O-ring problem and its decision to permit the launch of the *Challenger* violate business ethics? Michael Josephson, of the Josephson Institute for the Advancement of Ethics, reminds us that "very often people do overstate risks. An engineer will usually overemphasize risks, and a manager will use that safety margin to be less cautious." Why? "The manager is ultimately accountable," replies Josephson.

Practically speaking, the booster only had to fly for two minutes. If the booster had blown its seal just 50 seconds later, the *Challenger* would have been separated by that time and the only people to realize something had gone wrong would have been NASA and Thiokol engineers and managers. A defensible gamble?

One might argue that to the extent the space shuttle program was seriously underfunded from the outset, we all share a little of the blame. But Rex Finke, senior staff member for the Institute of Defense Analysis, has studied the budget in depth and argues that NASA got almost the exact amount that it asked for at the onset of the program. That sum was $5.25 billion, in 1971 dollars, he points out. Congress gave NASA $5.15 billion in 1972. Alex McCool, NASA's head of safety and quality assurance at Marshall Space Flight Center, remarks, "I have been here since the days of Von Braun. Even when we started to look at the system in 1972, we went with the solid rocket motors because of cost."

Finke confirms that the original plan for the shuttle system was to use liquid fuel boosters instead of solid fuel. But largely for cost reasons, NASA changed the plans and went with solid fuel boosters.

This is important because the liquid boosters are safer in some important respects. Tom Mobley, manager for shuttle applications at Martin Marietta, explains that there are several advantages to using liquid fuel boosters instead of solid fuel. "The liquids have a much higher payload capacity and the capability of being shut off if a problem occurs," he says. "Once the solid rocket boosters are turned on, there is no way to stop them from firing."

Trouble is, liquid boost cost 25% to 50% more.

One can also argue that the unrealistic charade of competitive bidding is to blame. Morton Thiokol won the booster contract because they bid lower than the competitors. In a Dec. 12, 1973 report, NASA selection officials said Thiokol's "cost advantages were substantial and consistent throughout all areas evaluated." Ironically, the report singled out the O-ring joint design for special praise for its cost-effective use of materials. According to one engineer, the steel that is used for the massive booster cases is considered the cheapest possible that will withstand the stress.

Redesigning the boosters has cost NASA more than $500 million, just over 60% of the original estimated cost of the booster contract itself. Was it reasonable to expect Thiokol to impose unilaterally upon itself such massive cost overruns and risk losing all future contracts?

Meanwhile, not one of the Thiokol managers who made the decision to launch was fired. And in return for a $10 million reduction in award fees, Thiokol was exonerated by NASA from any legal blame for the accident.

WHISTLEBLOWER[4]

The Challenger *disaster claimed more than the lives of seven astronauts; it was a malfunction that punctured NASA's—and America's— aura of high-tech invincibility. Roger Boisjoly, an aerospace engineer since 1960 and a 20-year veteran of NASA projects, worked on the O-ring*

[4]Interview conducted by Tony Chiu with former Morton Thiokol engineer Roger Boisjoly. *Life*, 11:17+. Mr '88. Copyright © Tony Chiu, 1988. Reprinted with permission.

*seals meant to safeguard solid rocket booster joints. In 1985 he warned his
employer, Morton Thiokol, Inc., the Utah builder of the* Challenger's
*rockets, of flaws in the seals. The government-appointed Rogers commis-
sion, which probed the tragedy, concluded its cause was leaking O-rings.*

*Diagnosed as suffering traumatic stress, and reviled within Thiokol,
Boisjoly (pronounced* BO-je-lay) *left his $50,000-plus job in January
1987. He has since filed two suits against MTI. One seeks $1 million in
personal injuries. The other, in behalf of the U.S. government, asks for
$2 billion, the value of the destroyed shuttle. His reason for bringing the
second suit was to establish that technology managements will be severely
penalized if they suppress advice from their engineers. For his campaign
to promote integrity in industrial decision making, Boisjoly last month
was awarded the prestigious Prize for Scientific Freedom and Responsibility from the American Association for the Advancement of Science.*

*Boisjoly contends in this interview with Tony Chiu that without a
thorough redesign of both the craft and NASA, the shuttle remains an accident waiting to happen—again.*

On January 27, 1986—the day before *Challenger's* **launch—you
learned disquieting news.**
That afternoon a NASA manager called from Cape Canaveral to
say meteorologists were predicting an overnight low of 18° F. I
went right into orbit, just absolutely came unglued, and started
to raise hell.

I was Thiokol's lead engineer on solid rocket motor joints.
The year before, in slightly warmer temperatures, we experienced massive primary seal blowby [gases escaping the innermost
of the two rubbery O-ring seals at each joint]. I felt cold weather
had been responsible—and here we had a condition way outside
what we ever expected.

**That evening you joined 33 other engineers and administrators
in a teleconference linking Thiokol in Utah with NASA at both
the Cape and at Marshall Space Flight Center in Alabama. The
topic: cold weather. Was a telecon on the eve of a launch
unusual?**
To my knowledge, it was the first in 25 shuttle missions.

The session was heated and lasted almost two hours.
That's normal for flight industry reviews. The very nature of the
system is to prove it's safe to fly. That night, those of us concerned with the temperature were challenged: Where is your
proof? Where is your data? That kind of probing is acceptable.

Some of the things NASA booster manager Larry Mulloy said—like, "My God, Thiokol, when do you want me to fly? Next April?"—went beyond probing; it was the start of intimidation.

But even with that, our chief engineer said he would not recommend launching. The data were inconclusive, meaning you must default to no-launch. Period, end of discussion. The deputy director of Marshall said he was appalled by Thiokol's decision but would not go against it. These two noes are significant. They remove any doubts about miscommunication—the managers fully understood the ramifications of launching at that time.

After the no-launch decision, one of your vice presidents requested a private Thiokol caucus.
When the telecon was put on hold and the senior MTI guy in the room said we had to make "a managerial decision," I knew we were in deep yogurt. Two of us tried to present our data again— by now I was so pissed my voice had gone up two or three octaves—but all we got were dirty looks from MTI management. They perceived NASA wanted a launch, so they began searching the data for interpretations to support a launch. It took them just 15 minutes to reverse the recommendation.

After the meeting I went to my office and wrote a heated entry in my notebook. It was way after hours, so three of us rode home in a company car. We had a very angry exchange all the way. I didn't get to sleep until real, real late.

Describe the next morning.
When I got up, I knew I wasn't going to watch that launch.

Didn't you usually?
Yes. It's a thrill to see a nice piece of machinery go up. But I didn't want to be an observer of something I was scared to death would have a problem. As an engineer I felt we had to have a few things go *right* in order not to have a failure.

I went to the eight a.m. meeting of the seal task force team, then stopped outside [fellow MTI engineer] Arnie Thompson's office and said I hoped the launch was successful. I added in disgust that I hoped the seals burned almost all the way through. Then maybe somebody would have enough guts to stop the launches until we fixed the joints.

I was walking by the conference room where we watched launches on a large projection TV when Bob Ebeling [another MTI engineer] came out and coaxed me inside. The room was filled with maybe 50 or 60 people—it was only minutes to go—so I ended up sitting on the floor up front.

What were you thinking about during the final countdown?
The propellant experts had predicted that if we had a leaking seal, the sucker would blow up on ignition. That's why, when we achieved liftoff, Bob whispered in my ear, "Hey, man, we just dodged a bullet." I began to feel very relieved. It was pleasure to see something pulled off in the face of such adversity.

At the one-minute mark, Bob leaned over again and said he had just completed a prayer of thanks for a successful launch. He was settling back in his chair when the thing splattered all over the sky.

Visitors in the room who'd never seen a launch thought it was separation. But the rest of us knew, instantly. We all sat stunned. After three or four minutes, I got up and went straight to my office. I sat in my chair, feet up on the desk, staring at the wall, fighting tears.

Later, two colleagues stopped in to ask some questions. I couldn't answer. Nobody else came by for the rest of the day, nor did I go talk to anybody—I just couldn't. At the end of the shift I got in the pool van. Didn't say a word, just put my hands on my head and rode home.

When was O-ring erosion first noticed?
The second shuttle flight [in November 1981]. But the problem preceded my arrival at Thiokol. There is a NASA memo from 1979 rejecting the design of the joint. Management at NASA and Thiokol ignored the flag because it would have meant a huge hit in costs. The lack of proper development work showed up when problems began surfacing. For instance, if everything works properly, hot gas should never reach the primary O-ring seal. But we had seal failures on nine of the 10 flights preceding *Challenger*.

It got bad, super bad, with the discovery of the massive blow-by on the January 1985 launch. Then when we opened up an April flight and found the seal in the fuel nozzle ring at the rocket's base to be totally shot, I started having sleepless nights.

What were you doing to alert Thiokol management?
Stating the problems in my weekly activity reports. Writing memos, some of which the Rogers commission published. You have to understand, I had never written a cover-your-ass paper in my career. But I was absolutely scared to death of what might happen if we didn't fix the joint in a timely manner. In July 1985 I wrote an extremely forceful memo warning of catastrophe. It was triggered by the memory of a colleague who worked on the

DC-10 cargo bay door. He saw a design flaw and tried to get the company to fix it. Instead, they went ahead and got the plane certified for flight. When the faulty door caused a crash in Europe that killed about 350 people, the designer held himself personally responsible for not doing enough.

The Rogers commission revealed the O-ring problem to be a lot more serious than you were aware of. It reported that in 1982 NASA reclassified the seals from Criticality 1-Redundant to Criticality 1. What's the significance?

There are two O-rings at each joint. Criticality 1-Redundant means the outermost, or secondary, ring is considered the insurance seal in case the primary fails. Criticality 1 means NASA considers the secondary ring nonexistent—if the primary ring fails, that's it. Morton Thiokol management never told us of the reclassification. The working troops at MTI understood we had a 1-Redundant. After the January 1985 blowby, we never would have tried to build a case for flying the shuttle, knowing there was no backup to the primary O-ring.

Why do you think NASA was so eager to launch *Challenger* **in the face of possible disaster?**

In my opinion there are four scenarios that have merit, either singly or in combination.

First, something might have come down from the Cabinet level because of a tie-in to President Reagan's State of the Union address that evening, at which he was to have discussed our future in space.

Second was the media ridicule of NASA's inability to get shuttles off on time. The agency insists this had no effect on the decision to launch. I think that's less than candid.

Third, NASA had to orbit a lot of time-critical payloads within specific launch windows. A delay would have had a domino effect.

Finally, Vandenberg Air Force Base was coming on-line in California. The temperatures there are colder. NASA may have wanted to use *Challenger* to show they could launch under any condition.

Had *Challenger* **not failed, would the O-rings have claimed a later shuttle?**

In a cool-weather environment, absolutely. After the disaster they ran subscale tests on the joint, using two different test platforms, at Marshall and Thiokol. The data show O-rings always fail at 25° F.

How much does an O-ring cost?
Roughly $900.
Is it fair to say that $1,800 worth of malfunctioning O-rings resulted in the loss of seven lives and $2 billion in hardware?
That's an oversimplification in that most failures usually occur because some minor subsystem gives: 25-cent washers, $2.50 bolts, $25 clevis pins. But yes, in the sense that if you do not take the time to develop and engineer and categorize the details, the details jump up and bite you.
You testified before the Rogers commission. Did its investigation satisfy you?
To be honest, I have not been able to bring myself to read anything but Volume One [the summary]. I think, though, they stopped short in a lot of areas. If you only look into the *Challenger*, you won't find out what else is wrong. Maybe that's all they were asked to do.
How did MTI treat you after you testified?
People I respected and trusted told me I would be a major player in the redesign effort. But by Mid-April I sensed that I was being isolated from NASA. When I made a technical presentation at Marshall, for the first time in six years nobody questioned a single thing I said. Just total silence. It was devastating.
How about your peers?
The five of us who testified called ourselves the lepers. There was a tremendous morale breakdown at MTI. We were getting blamed for it, though it was the company itself that pushed the self-destruct button. One guy actually went up to a colleague who testified and said, "You son of a bitch, you're ruining this company—and if it goes down the drain, I'm going to drop my kids on your front doorstep."
Have you been similarly ostracized away from the office?
Yes. Almost everybody in Willard [Utah], where my wife, Roberta, and I live, works for Thiokol. Nothing overt, but it's very real and it's deleterious. We plan to move.
You've sued MTI in behalf of the U.S. government. The False Claims Act allows the Justice Department to intercede. Do you expect it to?
I'm told the [FBI] investigation into Thiokol is ongoing, but Justice has had other aerospace contractors dead to rights without prosecuting.

Are engineers being ignored in industries other than aerospace?

Certainly. Look at the original Ford Pinto. The technical force pinpointed an engineering deficiency—rear axle gears punctured the gas tank when the car was hit from the rear. It's my understanding that it would have cost something like $14 a vehicle to fix the situation. Apparently, a cost-analysis study showed it was cheaper to defend lawsuits than make the fix.

Nuclear industry engineers are in a similar bind. The design work in general has been very fine. But I think fabrication with deficient materials, cheating on the welding and things like that have put them in the incredible position of not knowing what they actually have.

Since *Challenger*, **NASA has lost an Atlas and a Delta rocket on launch. Did the odds catch up with the agency?**

Perhaps, but I think it's more NASA's arrogance. They launched the Atlas in a lightning storm. All they would say later is that they did not violate launch-commit criteria. Hell, didn't anybody look out the window before pressing the button?

The agency is also trying to do too much with too little. Down at the cape, the shuttle assembly crews often worked horrendous overtime. You cannot work people in critical applications excessive hours day after day without the human being paying a price. On one flight, a technician accidentally dumped something like 18,000 pounds of fuel. Had it not been caught, the shuttle would never have reached orbit.

NASA was once held in high esteem. How has the agency changed in your 20 years of working with it?

Its can-do attitude is broken right now, but NASA still has an extremely gifted working-level staff and a lot of middle management people who are very competent. I think, however, it is virtually impossible to change the mind-set of a hierarchy that has been in place too long. Even without the big budgets that were available in the '60s, during the Apollo program, NASA management continues to play the game of meeting schedules at any cost.

Are you heartened by the return of longtime NASA administrator James C. Fletcher?

No. It's an exercise in damage control. Fletcher and Dale Myers, his deputy, were involved in some of the highly arbitrary decisions made during the original shuttle contract-writing process in the mid-'70s.

I hear NASA's having a hard time convincing Congress the space station is viable in its present form. I hope that's true, because from where I sit the agency is following the same principles that got it into trouble on the shuttle: namely, endearing themselves with low up-front cost estimates—even when it means accepting lesser technical design parameters for the initial phases. But when things get out of control downstream, they simply take more shortcuts instead of doing it right.

For instance, the computers on board the shuttle are mid-'60s technology, an absolute joke. People servicing them say spare parts are no longer manufactured—when they need new components, they have to make a special batch run. Very expensive. But NASA is willing to do that to show lower up-front costs.

Computer programs for in-flight operations and for astronaut training are no better. A Rockwell engineer, Sylvia Robins, says software changes have been installed without verification. One month there was an 18-inch stack of critical problems to be resolved; next month all that paper disappeared—there were "no problems." She has filed a suit like mine to try to protect the astronauts.

You've criticized the redesign of the nozzle joint, whose seals have also experienced erosion.
Yes. They redesigned the joint with 100 bolts, but the bolts penetrate the space between the O-ring seals. This creates 100 new potential leak paths if hot gas gets past the primary seal. The new joint ran into problems in a test at Marshall that's never been publicly reported and in a full-scale test at Thiokol in December. Both times, gas leaked through the joint, stopping just short of the O-rings.

NASA also has a continuing problem with the joint's boot ring. The first modification, even though it tested successfully last August, was not strong enough, and the agency was already developing another version, based on the Titan design. This version failed in December.

Incredibly, NASA is now reinstalling the model that worked in August. If you make the technical decision to redesign a piece of hardware, it is not on a whim—it is because something is wrong. And if the redesign fails, then you cannot ethically revert to the previous version and call it acceptable for flight. I'm afraid there is justification for saying that NASA has outlived its usefulness—to fly something that is less than right just shows the agency is still operating in the same pre-*Challenger* mode.

President Reagan proposed last month that certain space agency projects be shifted to the private sector. Is this a good idea?
As long as NASA remains the interface with the civilian contractors, it won't work.

Two years post-*Challenger*, what continues to motivate you?
A former chief engineer for Rocketdyne, which manufactures the main shuttle engine, said to me, "We make this immensely complex liquid-engine fuel-line system that is a nightmare of rotating machinery, advanced material, welding techniques, ad infinitum. I just can't understand how your lousy piece of pipe blew up."

Well, I think I do. If those with information don't bring it forth, how are you ever going to stop accidents from occurring again? I am not a loose cannon trying to get even with anybody—I keep speaking out because the space program is very important. I just can't stand to see crummy engineering.

Say somebody's approaching the edge of a cliff. Would you whisper—or would you scream?

II. THE TROUBLE WITH NASA

EDITOR'S INTRODUCTION

Since Challenger, NASA has undergone one of the most extensive reviews ever endured by a government agency. Many NASA managers associated with the disaster have been replaced, and the agency has had several changes at the top in the last four years. Morale is at an all time low, making it difficult to recruit and retain the best talent. The decline in the quality of NASA's staff and the reduced funding for existing programs have resulted in a string of disheartening technical failures. In 1990 the shuttles were grounded repeatedly by fuel leaks that NASA engineers seemingly couldn't fix. These were only the latest in a string of technical glitches that have reduced the number of shuttle flights per year from an early projection of twenty-five to a half dozen or less. Problems with the Hubble Space Telescope are probably the most embarrassing. "Hubble's Flaw Pinpointed," reprinted here from *Sky and Telescope*, explains in detail the problems with the telescope. Crippling flaws in the telescope's main mirror were discovered only after the two billion dollar observatory was launched in the early summer of 1990. In order to save money, NASA later revealed that during the construction of Hubble, it was decided to forego standard optical tests that would have uncovered the flaws.

The influence of money (or lack thereof) on NASA's mission is also at the heart of the debate over the proposed space station Freedom, the centerpiece of the nation's future space endeavors. In the journal *Science*, Eliot Marshall notes that NASA has revamped the space station concept repeatedly in response to Congressional budget plans. What is being cut, however, are the scientific projects that NASA originally advanced as the reasons for building the space station in the first place.

A more fundamental problem with America's civilian space program is its close ties to the Pentagon. Not only has NASA adopted the lowest-bidder, outside-procurement system developed by the Defense Department—with disastrous results in the case of Challenger—but the military was for years NASA's main customer, setting design and flight specifications that were often at odds with the purely scientific goals. For the shuttle program, this connection ended only in late 1990, when classified defense payloads were dropped from shuttle flights—not because of public pressure, but because NASA could not maintain a launch schedule that met the Pentagon's needs. Although the Defense Department is now developing its own launch capability, the damage may already be done. In the fourth selection, a 1987 article for the *Bulletin of the Atomic Scientists*, California Congressman George E. Brown Jr. argues that the military is diverting space dollars from peaceful exploration to dubious programs like the Strategic Defense Initiative, commonly known as Star Wars.

In the final selection, space policy analysts John M. Logsdon and Ray A. Williamson, writing in *Scientific American*, take a broad view of how launch-vehicle technology decisions have determined space policy, rather than the other way around, to the detriment of the space program as a whole. The writers argue that we need to know why we are in space and what we want to accomplish before NASA can do its job.

HUBBLE'S FLAW PINPOINTED[1]

Investigators have conclusively identified the sole cause of the Hubble Space Telescope's fuzzy images. The spacing between two optical elements in a device used to test the primary mirror's surface accuracy was discovered to be about 1.3 millimeters too large. As a result, the device was indicating that the 2.4-meter-wide mirror was perfect when in fact it had the wrong shape. The 1.3-m error completely accounts for the spherical aberration from which the telescope presently suffers.

Suspicion had narrowed to the test unit, known as a reflective null corrector, by last August 9th. But it took more tests and analyses for the NASA review board led by Lew Allen (Jet Propulsion Laboratory) to confirm that there was indeed only one problem with it.

The spacing between the affected lens and mirror was set with a metering rod, in effect a high-tech yardstick. Light was bounced off one end and into an interferometer, while the other end marked the location of the lens. The rod was then slid back and forth until it signaled (by the interference pattern it generated) that the lens was in the right place. To help aim the light, the rod's exposed face was covered by a cap with a pinhole at its center. On September 13th Allen announced that, by accident, the light beam apparently had bounced off a shiny spot on this cap, 1.3 mm in front of the rod itself.

Ironically, opticians at the Perkin-Elmer Corp. (now Hughes Danbury Optical Systems, Inc.) had information that could have allowed them to correct the mistake before the telescope was assembled and launched. After polishing the mirror, they checked the radius of curvature of its central zones using a second null corrector made entirely of lenses. It revealed clear evidence of spherical aberration, but apparently nobody at the facility did anything about it.

Review board member John D. Mangus (NASA-Goddard Space Flight Center) told *Sly & Telescope* that it is too early to jump to conclusions about this chain of events. The refractive null corrector was never intended to check the mirror's ultraprecise final figure, so it was not manufactured to the same high standards as the reflective unit. Consequently, Mangus explained, opticians may not have felt inclined to put much faith in its hint of a misshapen mirror.

To learn more, Allen, Mangus, and their colleagues began sifting through mountains of paper looking for documents that tell just how good the refractive null corrector was and thus how far it should have been trusted. They also began searching for anything to indicate whether Perkin-Elmer personnel ever told NASA of the questionable test result. C. Robert O'Dell (Rice University), HST project scientist for NASA when the mirror was manufactured, says he never heard about it. Neither, he adds, did two other astronomers helping NASA to monitor Perkin-Elmer's progress.

Meanwhile, as the investigation entered its third month, the people operating the spacecraft no longer had to be completely preoccupied with deducing the shape of the primary mirror from an exhaustive analysis of blurry images. They could instead rely partly on careful measurements of the flawed null corrector and thus turn some of their attention to other matters.

For instance, the telescope's two camera teams began making so-called science-assessment and early-release observations. The former, images of diverse examples of celestial objects, would show what kinds of research astronomers could still do—with help from computerized image restoration—despite the focus problem. The latter, "pretty pictures" of especially interesting and eye-catching scenes, would help restore the image of the telescope itself among the public.

Thus after two months of frustration and embarrassment, the mood at NASA began to take a positive swing. For project personnel, emphasis in HST's continuing orbital check-out changed from diagnosing how bad the telescope is to learning how good it is.

SPACE STATION SCIENCE: UP IN THE AIR[2]

This fall, Congress signed off on $1.8 billion for the space station—a 1-year installment on a total bill that, with shipping included, could run to $30 billion. But nobody—including members of Congress who voted the money, National Aeronautics and Space Administration (NASA) engineers who are designing the hardware, and scientists who hope to hitch a ride for their experiments—knows exactly what all that money will buy. Even at this late stage, 5 years after it was given the go-ahead, the space station has yet to be fully defined and its schedule completely worked out.

Since being included in the federal budget in 1984, the program has run through four directors, four agency chiefs, and 11 planning reviews, spending more than $2 billion along the way. And that was in the slow days. In the past 3 months, NASA has dismantled plans for the station and put them together again—not once, but twice. In addition, NASA is now overhauling the management. All this churning has upset scientists, who have

[2]Article by Eliot Marshall. *Science.* 246:1110+. D 1 '89. Copyright © 1989 by the American Association for the Advancement of Science. Reprinted with permission.

seen key scientific capabilities dropped, downgraded, or deferred, and it has worried NASA's international partners, who have feared they were getting shortchanged.

Even some of the space station's strongest supporters on Capitol Hill are getting concerned. Confusion about what the space station will do and when it will do it got Representative Robert Roe (D-NJ) angry. And when he gets angry he says so. As chairman of the House Science, Space, and Technology Committee, he called the program chiefs in for a dressing-down on 31 October, complaining that it is too late in the day to be changing the fundamental nature of the project, as NASA had done between August and October.

Roe warned that NASA will have to freeze the design and live with the consequences next year, for if the situation doesn't improve in 1990, he said, his committee has "the guts" to ask Congress to put it out of its misery. Representative F. James Sensenbrenner (R-NY) agreed.

"What annoyed the committee," Roe told *Science* later, "was that they went ahead and reshaped the station to meet levels of budget allocation," rather than holding firm and asking Congress to vote an honest yea or nay on the original plan. "I wanted to make it clear," Roe said, "that . . . if you're going to reshape and reform every time we turn around, then we don't really have a U.S. space station; we have a station that's driven by funding rather than technology."

Roe reports that new delays in the schedule could add $200 million to $300 million to the cost borne by the European partners alone. And another committee member, Representative Robert Torricelli (D-NJ), told the NASA chiefs that by trimming the station's capabilities, "You are [turning] what presented itself as possibly the world's most advanced scientific laboratory into a giant orbiting recreational vehicle."

NASA's propensity to keep going back to the drawing board began to worry the foreign partners in August, especially because they were excluded from a review (the "rephasing") at which major cutbacks were debated. They feared they were losing out on power supply, communications links, and priority of hookup. Even for NASA to consider such changes unilaterally, they protested, was a violation of international agreements.

The furor may have affected the funding debate this fall. The House, which had been considering a large, $400-million cut in

NASA's request, relented in October and voted to take away only $250 million, leaving the station a total of $1.8 billion for 1990. In response to the protest, NASA also began restoring some of the items it had threatened to cut out, but not all of them. With more budget shuffling on the horizon (the Gramm-Rudman-Hollings process threatens to take back over $50 million of the money Congress just promised), even NASA people find it hard to see exactly what the station will include.

The plan is a blur, changing almost every week, like "a train roaring through my office," says Robert Rhome, a NASA science official whose job is to ensure that researchers get a berth on the train as it flies past. For space science, there are some big unresolved issues. Rhome laid them out in a presentation to a NASA control board in Reston, Virginia, on 19 September. There is not enough power, he said. The data management system is scheduled to grow too slowly. Not enough crew members will be available to run experiments. In its initial configuration, Rhome said, the station did not seem to be much of an improvement over the shuttle's spacelab as a research center.

The Office of Space Science and Applications, for which Rhome works and which is headed by Lennard Fisk, will soon begin to negotiate with the space station program, headed by Richard Kohrs, on exactly who will pay for what. Within the next 8 months the fundamental issues must be settled. Starting in July, the program will undergo a "preliminary design review" when most of the fuzzy lines in today's concept must be made hard.

Already, the funding for some lab equipment has been postponed beyond 1991 and other items have been shoved beyond NASA's "budget horizon" of 1995. This approach "negatively impacts life sciences research," says Laurence Young, chief of the man-machine laboratory at the Massachusetts Institute of Technology and head of a NASA biomedical advisory group.

Young is disappointed that the new schedule has cut the crew size from eight to four through 1997, for this means crew members will be so busy with routine chores that they will have little time for science.

NASA thinks this concern may be overstated. Responding to questions from Representative Roe's committee in November, NASA reported that the station crew will spend 240 hours on scientific investigations in 1996, 2,600 hours in 1997, 5,000 hours in 1998, 11,000 hours in 1999, and 15,000 hours in 2000.

Young remains concerned, and , in fact, has another worry. NASA has not yet budgeted for $200 to $300 million worth of research hardware. One important item—a centrifuge for biomedical studies—is slated for early deployment, but other tools may not be.

For example, according to NASA's Rhome, the original design included racks of freezers and other equipment to hold specimens awaiting shipment back to Earth. Now they are gone. Young says, "The presumption is that [the cost of installing the freezers] will be picked up again" by NASA's science office. But Fisk has not yet offered to pay, and Kohrs said recently that he found the list of things Fisk's office wants him to do "not all that well thought out." But he insists that "NASA will pay," one way or another. The risk, according to Young, is that decisions will be put off until it is too late to shoehorn everything that's needed into the schedule.

It would be hard to do good research without freezers, for example, Young says. Experiments would have to be coordinated with shuttle visits so that specimens could be loaded quickly into the cargo bay just as the shuttle prepared to go back to Earth. Such rigid scheduling—in addition to the many other chores that must be done during shuttle stopovers—might be too demanding.

Equally important, according to Young and Rhome, is a system to carry live animals from Earth to the station and move them quickly from the shuttle to the orbiting lab. At the moment there is no provision for the $24-million "animal specimen transport system." But, as Fisk wisecracked recently: "The station won't be of much use if we have to use freeze-dried animals." Fisk has offered to begin paying for a system in 1992, not before. If funding is allowed to slip too long, Young warns, "the whole space biology program will disappear."

Materials scientists have a list of problems of their own, reflected in comments submitted to an advisory group by Robert Bayuzick, a materials scientist at Vanderbilt University, and Simon Ostrach of Case Western Reserve. Their greatest concern is that the station be equipped with gadgets to monitor and control forces of acceleration throughout the structure, specifically to ensure that experiments requiring microgravity (one-millionth of Earth's gravity) can be sustained over a 30-day period. This is "an absolute necessity," according to Bayuzick, one of several that NASA has not yet agreed to provide.

NASA administrator Richard Truly told the Roe committee that the space station will be "virtually the same functionally as that envisioned in the [original] program," except for an 18-month delay in its completion. The first parts are slated to be launched in 1995 and the crew is to begin living aboard in 1997. Truly's reassurances had a calming effect, but the foreign partners remain uneasy.

Takehiko Kato, Japan's liaison officer to the space station program, says the decision to cut the station crew back from eight to four will create many problems. He also dislikes a recent decision to substitute the toxic chemical, hydrazine, as a thruster jet fuel on the station rather than using hydrogen and oxygen. By adopting an already developed hydrazine technology, the United States lowers its development cost, Kato says, but increases the operational cost over the long term. Hydrazine supplies will have to be shipped to space on a regular basis (whereas hydrogen and oxygen could have been generated by equipment on board). This will require extra shuttle flights, the cost of which must be borne by all the partners, as must all increases in the operating budget. The Japanese Diet may see that as an unbargained-for new expense.

Kato also sees a safety risk: the astronauts' space suits could get contaminated with hydrazine while they are working outside the station. He wonders if it will be necessary to add a special suit scrubbing facility.

But Kato has a deeper concern. He sees "instability" in the program's management and harbors doubts about Congress's commitment to future funding. While "the main issues" that caused upset earlier this fall "are resolved," he says, "we will have to wait and see how the new management does."

Kato's counterpart at the European Space Agency, Derek Deil, says his "concerns are exactly the same" as they were earlier this year, but "we are pleased to see that NASA is trying to work things out now."One awkward proposal was scrapped. NASA wanted to attach the European laboratory module a year before turning on full power for research. "It would be very difficult to explain to our politicians that we are taking up a billion-dollar lab but will not be able to operate it until a year later." Now the plan is to get the power on first. "The mood has certainly improved" since last August, Deil says.

Karl Doetsch, director of Canada's space station program, says the allocation of power is an unresolved issue "of great concern." So is the availability of data processing, communications links, and robot systems that can be used to relieve labor demands on the crew. Canada is supplying the Mobile Servicing Center, a huge ($1.2-billion) mechanical arm that will be critical in assembling the station. Doetsch warns that NASA's plan to reduce the crew from eight to four may require a bigger investment in automated machinery, which the agency does not now plan to make.

The reduction in crew came about through a cascading series of cutbacks, according to one NASA official. When the planners decided to delay installation of a complex and heavy oxygen regeneration system, they realized that the task of supplying eight people in space with bottled oxygen would be horrific, turning the station into a "hungry beast," devouring many shuttle flights simply in the task of feeding its inhabitants. The best solution, it seemed, was to reduce the crew to half the number until the construction phase was near an end.

At present, NASA plans to reach the point of "assembly complete" in late 1999. By that time, the crew is scheduled to grow to eight and other items that have been dropped temporarily will be put back. However, Doetsch, like several others, remains troubled by "the complete lack of definition of when we reach the 'assembly complete' configuration."

Another ill-defined point that concerns the Europeans is the vague plan to convert the station into a "transportation node" for U.S. trips to the moon and beyond. President Bush has said that Americans will build a lunar base early in the next century. Some worry that if this goal becomes a reality, researchers will enjoy a relatively brief "quiet period" on the station before the construction of the lunar base begins. However, a NASA official involved in planning for the future, Frank Martin, dismisses these worries, saying, "We're not going to send people with jackhammers up there."

The moon mission and some other items on the agenda remain fuzzy, which has been a problem for NASA's planners in the past. For example, Fisk, as he met with his outside advisers (the Space Science Applications Advisory Committee) on 9 November, reflected that trying to agree on items to be included in the station "has been a very frustrating experience" because the responses to requests were often vague. "If someone says, 'No,' then you can appeal it," Fisk explained. "But if they say, 'We're working on it,' what do you do?" He told the group that the outlook is getting better because the new managers don't like to leave things unresolved.

Some of this agitation—such as the demand for first-class passage for lab rats—may sound trivial to NASA engineers who are trying to put the structure in orbit and keep its crew alive. "Scientists are traditionally guilty of special pleading," says Radford Byerly, Jr., an expert on space policy who once headed the staff of the House space science and applications subcommittee and is now director of space policy studies at the University of Colorado at Boulder. "On the other hand," Byerly says, "NASA has advertised the space station as a research laboratory" and invites judgment on that basis.

Byerly and his colleague at Boulder, Ronald Brunner, dissected the space station in a sharply critical paper last month, finding that the root cause of its problems is a lack of "resilience." They argue that the program was conceived with a bureaucratic aim—increasing the size of the space program—and cast in a form that is huge, complex, indivisible, and very difficult to steer. As a result, Byerly and Brunner say, the program keeps crashing into fiscal barriers and being forced to redefine itself. With each redefinition, some of the earlier promises are postponed. This gives the impression that money is being wasted, leading Congress to impose tighter fiscal controls, triggering self-review exercises like the one this fall, and more deferrals. The only way for NASA to escape the cycle, according to Byerly and Brunner, is to "decouple" the elements of the space station and redesign it as a series of smaller, independently viable projects. These projects should be ranked by priority, the critics say, and built as funds become available.

This is not the way NASA would solve the problem. At the hearing on 31 October, Richard Truly reiterated NASA's long-held view that Congress could improve the program by passing a multiyear funding bill. The assumption is that if NASA could go about its business without annual interruptions from Capitol Hill, it would do the job more efficiently.

There are some leaders outside NASA who think multiyear funding is a good idea—including the chairman and ranking member of the House science committee, Representatives Roe and Robert Walker (R-PA). But the idea probably won't get much support elsewhere in Congress, particularly not in the appropriations committees where it would count. Support for the space station is already tenuous. Given the prospect of continued tight budgets through the early 1990s, committee members are not likely to give up any of the power of the purse they still hold. For this reason, NASA is not likely to get the multiyear appropriation that it seeks, and the space station could face a series of annual reviews as harsh as this year's.

Optimists think differently, and there are many optimists among the space station's planners. They are convinced that 1990 will be the year in which the program straightens out. "When you're in the space business, you have to be optimistic," says Kato, Japan's liaison officer who seems to have absorbed the NASA esprit de corps while working in Reston,Virginia. "We have to live today before we can worry about tomorrow."

PENTAGON USURPS CIVILIAN SPACE PROGRAM[3]

Although people generally think of NASA when they think of the US space program, these days they should really think of the Pentagon. For although NASA has brought the United States the most spectacular accomplishments when it comes to outer space, NASA's efforts are now dwarfed by what the Pentagon is doing, and wants to do, in space.

Nearly 70 percent of the nation's overall space budget in recent years has been controlled by the Pentagon. This was not so a decade ago. During the 1970's, NASA held a modest edge over the Pentagon in funding. During the past six years however, funding for the military space program has skyrocketed. Nearly $80 billion has been spent on military space activities during the Reagan years. The rate of growth in these programs has been substantially greater than that of the rest of the administration's defense buildup.

[3]Article by Congressman George E. Brown, Jr., of California, ranking Democrat on the House Science, Space and Technology Committee and member of the House Permanent Select Committee on Intelligence. *Bulletin of the Atomic Scientist*, 43:26+, N '87. Copyright © 1987 by the Educational Foundation for Nuclear Science, 6042 South Kimbark Avenue, Chicago, IL 60637. Reprinted with permission.

The Pentagon's space budget surpassed NASA's for the first time in 1982. At that point, both programs had spending authority for roughly $6 billion. Since 1982, NASA's budget has barely kept up with inflation, while the Pentagon's space budget has more than tripled. The Pentagon will spend on the order of $20 billion in fiscal year 1988 on space-related activities. A review of these activities shows that the nation's military space effort is incredibly expansive, involving dozens of satellites conducting a multitude of missions.

Many of these missions, such as the verification of arms control agreements and the maintenance of deterrence, are enormously valuable to security. A few for the missions assigned to military spacecraft, however, are disturbing and relate to the misguided objective of developing a nuclear war-fighting capability. What becomes most apparent from surveying the military space program, though, is the manner in which the current administration (Reagan), in sharp contrast to its predecessors, is developing the means for engaging in space warfare while at the same time cutting back the nation's peaceful priorities in space.

Intelligence gathering is an important, legitimate function of the military space program. From their unique vantage in space, satellites can photograph inaccessible regions of the Soviet Union; they can eavesdrop on telephone calls, radio transmissions, and telemetry emitted during Soviet missile launches; they can also monitor troop movements and military conflicts anywhere in the world. These military satellites include:

• *Keyhole (KH) image-collecting satellites.* From the KH-1 (also called Corona) in the early 1960's to the KH-11 and the KH-12 in service today), these devices have been dramatically refined. the KH-11 reportedly can monitor an object the size of a paperback book from a distance of more than a hundred miles, and can transmit images of events underway anywhere almost instantly to the Oval Office.

• *SIGINT (signals intelligence) eavesdropping satellites.* Operated by the National Security agency, **SIGINT** can intercept everything from diplomatic communications to conversations between military officials or among political leaders. As many as nine **SIGINT**s orbit at a time, and they have uncovered terrorists plots, monitored Soviet military exercises, and listened to conversations between Oliver North and arms brokers in the Middle East.

• *early-warning satellites*. Designed to provide warning of a missile attack, these satellites can detect Soviet missile launches 30 seconds after they occur. The United States has three such satellites, and these would provide instant information on the speed and course of the approaching missiles; they are fundamental to the US nuclear deterrent.

• *military communication satellites*. These devices account for more than 20 of the approximately 75 US military satellites operating. They act as extremely tall relay towers for global communications and carry nearly 80 percent of the US military's long-distance communications traffic. The Defense Satellite Communications System, now the military's principal communications system, will by the early 1990s be augmented by the seven-satellite Milstar system. Milstar is being designed to be survivable against many forms of interference and attack. At a price tag of $10 billion, it is the most expensive satellite communications system ever conceived.

• *navigation satellites*. In use since 1959, navigation satellites of the Transit system have provided aid to military and civilian users through more than 35,000 receivers mounted on ships of all kinds. Transit will be replaced in the early 1990s by the Navstar Global Positioning System (GPS), the most sophisticated military satellite system ever. GPS will consist of 18 operational satellites and will allow military navigators to determine with unprecedented accuracy, their vessel's location and speed. Civilians will have limited access to the system but will receive substantially less accurate data. The cost for GPS through 1992 may exceed $6 billion.

But while GPS will provide valuable navigation information, the extent of its precision raises serious concerns. GPS is as good an example as any of how the military could, and indeed seems inclined to, greatly increase the targeting ability of nuclear weapons by using ostensibly peaceful elements of the US space program. The navy has already indicated its interest in equipping Trident II missiles with in-flight course-correction abilities afforded by GPS. This raises serious concerns since such accuracy is necessary only if one intends to strike first in a nuclear attack; it is not required to maintain retaliatory capability, which should be the United States' nuclear posture.

Of additional concern regarding GPS is the potential role of the nuclear-detection sensors for the spacecraft. These sensors, known as the Integrated Operational Nuclear Detections System

(IONDS), are designed to provide military planners with instantaneous information about nuclear explosions occurring during what the Reagan administration refers to as "protracted nuclear war." IONDS could detect the precise locations and effects of hundreds of simultaneous nuclear explosions. With these data, Pentagon planners expect to be able to assess which targets have been destroyed during a nuclear war, and which need to be hit again.

The plan to spend $700 million on a system designed to help wage a protracted nuclear war is alarming. There is no rational description of what protracted nuclear war is, nor any reasonable basis for preparing for such an event. Unfortunately, IONDS is simply one of the many pieces of hardware the Reagan administration is accumulating in its drive for a nuclear-war-fighting capability. Many of the most expensive space systems now under development are being designed to participate in space warfare. This reflects the vision of space now governing Pentagon thinking: no longer a medium to support satellites for passive communications and intelligence-gathering purposes, space is an emerging theater for combat operations and the new "high ground" awaiting domination and control.

The Reagan administration's space policy has been spelled out by Pentagon officials and documents throughout the past six years. For example, in 1982 Gen. Bernard Randolph, then director of space systems for the Pentagon, told the Senate Armed Services Committee: "We are working toward making sure that we have a fighting capability, a warfighting capability, with the space system." The air force *Military Space Doctrine* manual of October 1982 claims: "The Air Force will maintain U.S. technological superiority in the aerospace, and ensure a prolonged warfighting capability by developing the potential for combat operations in the space medium." Gen. Richard Henry, former deputy commander of the Air Force Space Command, called space "the new high ground of battle" in an article in the May 1983 issue of *Military Electronics*. And in testimony before a House subcommittee on defense, Robert Cooper, former director of Pentagon research, said the administration's space policy "for the first time really recognize[s] the need to be able to control space as a military environment. It directs the kinds of activities. . . . that may be necessary to mount space weapons in the future if that appears in the national interest."

These statements represent a drastic departure from the
views held by past U.S. presidents. Dwight D. Eisenhower was
deeply concerned about keeping outer space from turning into
an armed camp, and in a 1960 speech to the United Nations General
Assembly he warned of starting a "dangerous and sterile
competition" in outer space. In 1961 John F. Kennedy told the
United Nations: "The cold reaches of the universe must not become
the new arena of an even colder war."

Building on these sentiments, in 1967 Lyndon Johnson signed
the Outer Space Treaty with the Soviet Union, banning the deployment
of nuclear weapons in space and asserting that space
should be preserved for peaceful exploration and use by all humanity.
Richard Nixon added to the space arms control regime
with the 1972 Anti-Ballistic Missile (ABM) Treaty, which prohibits
the development and deployment of space-based ABMs or
their components. Jimmy Carter sought additional curbs on the
militarization of space by initiating U.S.-Soviet negotiations
aimed at limiting antisatellite (ASAT) weapons. Carter's goal for
these talks, according to fiscal a year 1979 Arms Control Impact
Statement, was to "prevent an arms competition in space and to
reduce the threat to U.S. satellites."

In the wake of these presidential efforts, Reagan's treatment
of space amounts to a complete policy reversal. The attempt to
redefine the ABM Treaty in order to permit the testing of exotic
weapons in space is but one example of the administration's disdain
for arms control measures governing space weapons. Another
example is its treatment of ASAT arms control. Rather than
seeking to preclude a U.S.-Soviet competition in these weapons,
the Reagan administration's principal objective has been the deployment,
as rapidly as possible, of a $4 billion ASAT force.

Arms control of any kind in space has been effectively rejected
by the Reagan administration. In a March 1984 report to Congress
on ASAT arms control, the administration made its
position clear: "No arrangements or agreements beyond those already
governing military activities in outer space have been
found to date that are judged to be in the overall interest of the
United States and its Allies." In other words the administration
has no interest whatsoever in preventing the militarization of
space. On the contrary, since the time of Reagan's March 1983
Star Wars speech it has been obvious that the principal future the
administration sees in space is one where exotic weapons are de-

ployed in massive numbers. The path to that future is being forged by the Strategic Defense Initiative (SDI).

We have heard a great deal from the (Reagan) administration about proceeding with the first phase of an SDI deployment. Although the precise composition of the system is unclear, the general outlines of what (former) Defense Secretary Caspar Weinberger and Lt. Gen. James Abrahamson, the director of SDI, have in mind are known.

An initial-phase SDI system apparently would consist of three layers of weapons: two layers of ground-launched missiles and a third layer of space-based battle stations. In one layer of ground-launched missiles, as many as 10,000 missiles designed to intercept targets as they travel through space are envisioned. Another layer consisting of some 3,000 smaller missiles would attempt to intercept nuclear warheads during their last few minutes of flight. The space-based layer would involve at least 2,000 battle stations, each equipped with five or more vehicles for attacking Soviet missiles and warheads in space. An array of sensor satellites would also be orbited. The entire system is projected to cost $100-$200 billion.

How effective this system would be is unclear. That would depend on the technologies involved and on the nature of the Soviet response. Whether the weapons proposed could meet the incredible technical demands of their mission is questionable, and there are strong indications that the Soviets will respond quickly to make the weapons obsolete. According to the Pentagon's annual report on Soviet military forces, the Soviet Union has an aggressive program underway to develop the same type of space weapons now suggested by the administration. Space-based laser systems, orbiting kinetic-kill battle stations, space-directed interceptor missiles—all of these are reportedly being developed by Soviet military researchers. The scenario that emerges is one in which both nations deploy successive layers of weapons in space and on the earth.

This measure-for-measure competition would bring a host of risks. Space would become a hostile arena. The weapons would be targeted against each other as well as against military satellites of all kinds. The command, control, and intelligence satellites mentioned above would become vulnerable to attack as never before. For this reason, both superpowers would seek to equip their satellites with the ability to shoot at other satellites. Indeed, the

Pentagon is already well along in developing satellites that could engage in orbital dogfights. Protecting a space-based defense from attack would be a serious problem. Any defense most likely would be deployed in orbiting armada formations: central battle stations would be surrounded by constellations of defensive layers, analogous to carrier task groups in which rings of surface ships, planes, and submarines are arrayed to defend aircraft carriers. Each action would call for a response; each response would invite a counter.

There would be no victor in such a competition, since neither nation would permit the other to develop mastery over space. And this competition would be dangerous and enormously expensive. At any point along the way, malfunctioning battle stations could precipitate an escalating crisis and perhaps trigger a nuclear war.

The military space program, as noted earlier, already towers high over the civilian space effort. This disparity would be exacerbated if SDI were deployed. The launch requirements of deploying a space-based defense would be staggering, threatening to consume an enormous portion, if not all, of the nation's launch capacity. According to some descriptions, deploying an initial SDI defense would require lifting 2.6 million pounds of weapons and sensors into space per year to begin with, increasing to 4.4 million pounds annually by the year 2000. This might require an SDI launch every other day, or 200 launches per year. This compares to an average U.S. launch rate over the past decade of fewer than 20 per year.

SDI's appetite for launch capacity would wield greater and greater control not only over the use of existing launch vehicles, but also over the decision-making process regarding future launch systems. This could cause further delays and setbacks to scientific and civilian space launches. In order to bring launch costs down, a whole family of new launch vehicles would be needed. The National Aerospace Plane, a new heavy-lift vehicle, and possible other boosters would be involved in getting an SDI system into space. Investments in additional launch systems alone could well exceed $50 billion.

Much more than new boosters would be required to expand launch capacity and lower launch costs to the levels necessary for SDI. SDI's transportation and operation needs would necessitate drastic changes in the nature of the U.S. space program. The complete mobilization of U.S. space resources could well be required for the single objective of deploying, operating, and maintaining a space-based defense. NASA could simply become a trucking company for SDI. In many ways, NASA already resembles a trucking firm for the Department of Defense: nearly 50 percent of future shuttle launches are slated for the Pentagon, and many of these missions will involve SDI tests.

NASA's proposed space station fares no better than the shuttle. Using the space station as an SDI laboratory is one interest of the Pentagon; using it as a service station for repairing military—including SDI—satellites is another. NASA would be pulled deeper into SDI's mantle if the program continued toward deployment, further jeopardizing NASA's status as an independent agency.

The administration will have spent more than twice as much on SDI during 1987 as it has this year on NASA's entire space science and applications program. It will have spent 100 percent more on the development of directed-energy weapons than it has on space-exploration missions. And it will have devoted substantially more resources to SDI kinetic-energy weapons than to all of what NASA is doing in physics and astronomy. These contrasts are revealing and startling, and they convey a resounding message about U.S. priorities for the future of space. With the Pentagon's space program being bankrolled while NASA continues its downward spiral, one can be sure that SDI and the Pentagon will harvest a bountiful crop of researchers and business executives who, if otherwise faced with a different set of national priorities, would have looked instead to NASA for work.

A late-1986 poll of 200 space scientists at Stanford University projected that 50 percent of the graduate students in civilian space research would soon leave the field to seek careers elsewhere. These scientists are fed up with the delays, cutbacks, and cancellations that have plagued NASA's space science programs over the past decade, and that have become especially acute since the *Challenger* accident. Many of these departing space scientists will naturally turn toward the Pentagon, which runs the nation's principal space program these days, and to SDI, which is the largest and best-financed research effort in history.

The thinking of these scientists is certainly understandable. Why should they agonize over the recent cancellation of 30 space science missions that had been scheduled aboard the shuttle, when they can participate instead in some of the administration's high-priority SDI tests? Or why should they care about the administration's failure to support NASA's efforts in earth remote sensing, in advanced communications technology, or in X-ray astronomy, when they can work instead on intelligence-gathering satellites for the Pentagon, next-generation military communications spacecraft, or nuclear-pumped X-ray laser devices—all of which are being lavishly funded?

Thomas Donahue, chairman of the National Academy of Sciences Space Science Board and one of the nation's most prominent civilian space scientists, said in November 1986: "I would consider myself to be irresponsible to advise students to come into this field [civilian space science] right now." Professors working on SDI and on other military space programs probably have the opposite perspective and are urging their students to jump aboard the military space juggernaut.

This is disturbing, not only because of the damage it will do to the long-term health of the nation's civilian space program, but because it represents a fundamental movement away from the peaceful uses of space and toward military uses of space that is affecting the nation's overall vision of its future in outer space. When NASA was striving to land a man on the moon, space was seen as the next frontier and astronauts assumed the role of modern-day pioneers. Today, the prevailing vision of space seems to involve orbiting battle stations and space warfare. Astronaut pioneers have been replaced by astronaut warriors in the minds and imaginations of our youths.

A space arms race can and should be prevented. Doing so, however, will require three things:

• *resuming the effort to expand the arms control regime in space.* This will require strengthening the ABM Treaty by more clearly defining what is and is not allowed in terms of research and development. The administration's effort to broaden the ABM Treaty to permit testing of exotic ABM systems in space should be rejected. The existing arms control regime in space should be supplemented with an ASAT treaty designed to strictly limit both superpowers' abilities to develop and test sophisticated ASAT systems.

• *developing a more deliberate process of setting U.S. priorities for space.* The Pentagon's growing control over the U.S. space program has resulted from the lack of public debate, and from the Pentagon's formidable ability to control national decision making and to direct national spending patterns. What is needed, at a fundamental level, is increased public discussion about what this nation should be doing in space during the coming few decades. Such discussions, catalyzed by politicians and the media, should compare, for example, the vision of the future developed by the National Commission on Space with that embodied by SDI.

• *pressing forward with a broad program of international cooperation in space.* Nothing would serve as a greater deterrent to a space arms race than a clear, compelling alternative, one involving nations throughout the world engaged in space science and exploration. These efforts could be aimed at various goals, including improving and internalizing the verification regime for arms control, enhancing our understanding of the earth by deploying an array of sensors to monitor environmental changes around the world, and expanding our knowledge of the solar system and the universe. It has been proposed that the United States and the Soviet Union work together on a number of space exploration missions in preparation for a U.S.-Soviet mission to Mars. A cooperative effort of such magnitude would represent a far greater contribution to the world's space experience than would a U.S.-Soviet competition in space weaponry.

III. BACK INTO ORBIT

EDITOR'S INTRODUCTION

In October 1988 the shuttle Discovery roared back into space, ending almost three years of paralysis of the space program. The near perfect flight was a vindication for NASA, which came under searing criticism in the wake of the Challenger explosion. Much of the criticism focused on flight preparation and readiness. "Shuttle Pit Stop," reprinted here from *Air & Space Smithsonian*, details the arduous process of preparing the shuttle for launch. It takes hundreds of people, months of preparation between one touchdown and the next liftoff.

Despite its problems, NASA still knows how to tap the source of American ingenuity. Innovative start up companies like Honeybee Robotics, profiled by Tom Waters in "The Robot's Reach" in *Discover*, provide the radical new ideas that a maturing organization like NASA cannot always generate from within. Tom Waters goes on to point out in "Moonstruck" that we have yet to answer the many scientific questions raised by the Apollo missions. President Reagan's decision in the mid-1980s to eliminate commercial payloads from NASA carriers created an opportunity for private corporations to get into space. Small American companies like Space Services, Inc. as well as major players like Martin Marietta are competing for commercial payloads with the nascent space programs of Japan, China and Europe. The result is what *Scientific American* writers Elizabeth Corcoran and Tim Beardsley call the new space race. The commercial space industry, still in its infancy, is already spawning new, low-cost approaches to space. Even NASA and the Pentagon are looking toward small carriers to reduce some of their own launch backlogs.

The space program is not just a matter of machines hurtling through a vacuum, it involves people who have the vision to design a mission and the expertise to direct it from millions, even billions of miles away. The final article in this section, reprinted from the *New York Times Magazine* is a profile of astronomer Ed Stone, chief scientist on Project Voyager. It was the engrossing pictures and discoveries of the two Voyager probes, which gathered more data on the solar system than had been collected in the previous four hundred years, that largely sustained the space program in the dark days after Challenger.

SHUTTLE PIT STOP[1]

In the pre-dawn darkness of California's Mojave Desert, the Orbiter Recovery Team huddles in front of an Airstream motor home loaded with sophisticated electronics. On a flip chart bathed in spotlights they review the procedures one last time for the task that is almost at hand. Three hundred miles above them and on the other side of the Earth, the crew of the space shuttle *Discovery*, which several days earlier had launched the Hubble Space Telescope, is preparing to somersault back home.

It is 4:30 a.m., April 29, 1990, on the dry lakebed of Edwards Air Force Base. As an assortment of vehicles rumbles across the desert, the *Discovery* crew prepares the winged spaceship for reentry. The recovery convoy will wait for the shuttle midway between the base's primary and alternate runways.

All together some 160 personnel are waiting here in the desert for *Discovery*. But they are only a fraction of the largest pit crew in the world. During the next five months more than 8,000 others will spend almost a million hours getting the world's only reusable manned space vehicle ready for its next launch.

The space shuttle was initially conceived as needing only two weeks of preparation between landing and launch. That proved overly optimistic. The quickest turnaround so far occurred when *Atlantis* was readied in just under eight weeks after a 1985 flight. Since the resumption of shuttle flights following the *Challenger* explosion in January 1986, the fastest turnaround has been 18 weeks.

[1]Article by Greg Freiherr, *Air & Space Smithsonian*. 5:70+, O/N '90. Copyright© 1990 by the Smithsonian Institution. Reprinted with permission.

This year's shuttle manifest called for a launch rate as high as the program had achieved just before the 1986 accident—nine per year, or one every five to six weeks. But last summer's problems with hydrogen fuel leaks briefly grounded the fleet after three launches. It also called the space shuttle's reliability into question.

It wasn't the first time. The *Challenger* investigation raised questions about whether the shuttle had been launched on an unseasonably cold day because of pressure to meet a schedule that was falling behind. In the post-*Challenger* era, NASA has stressed a new "safety over schedule" credo. Yet the pressures to get the shuttles back on the pad are always there.

"Sure there are timelines to meet," says Larry Ellis, deputy director of shuttle operations, about postflight operations. "I would call that pressure. It's a schedule . . . [But] if we didn't have a plan laid out it would just take as long as it takes, and we may never get there."

Time is irrefutably a luxury, especially in the moments immediately following a landing. After *Discovery* touches down, the spacecraft continues to roll for about a minute. When it comes to a stop, mission responsibility shifts from Houston's control room to Florida's Kennedy Space Center and its workers in the California desert.

While the shuttle crew completes their post-landing check, recovery team members approach *Discovery* and use sensors to ensure that nothing toxic is leaking from the orbiter. In the event of propellant leaks, a truck with a large fan stands ready to blow away dangerous gases.

The ground crew has only 45 minutes to hook purge and coolant lines to the tail of the orbiter. If they are tardy, the boilers that cool the shuttle's electronics run out of ammonia coolant and the electronics have to be shut down before they overheat. This would delay the flight back to Kennedy.

On a mobile staircase parked behind *Discovery*, technicians Tom Gay and Billy McClure wrestle with the 12-inch-diameter coolant hose. Suddenly the wind shifts and ammonia from the shuttle's boilers fills the air. To catch his breath, Gay has to step down from the staircase. "This isn't just household ammonia," he says. "This is 99 percent pure."

Gay and McClure normally work at Kennedy. To meet *Discovery* they have left work on *Columbia* and *Atlantis*, which are in different stages of the ground turnaround process. Here in the California desert they have to deal with wind gusts of 60 mph. Cars caught in these storms have had the paint blasted off them. For Gay and McClure, both used to the climate of the Sunshine State, these working conditions leave a lot to be desired. Gay succinctly sums up their feelings about the winds: "It's miserable."

During the next hour, other ground support equipment is connected to cool both the flight crew and the avionics while the recovery team readies the shuttle to leave the desert. Finally, their post-landing and systems checks complete, the *Discovery* crew—Loren Shriver, Charles Bolden, Steven Hawley, Bruce McCandless II, and Kathryn Sullivan—disembark down a staircase mounted on the back of the shuttle exit truck. This mission is over.

Almost a week later, after a flight back to Kennedy Space Center aboard NASA's modified 747, *Discovery* has been jacked about 10 feet in the air and encased in a skeleton of work platforms, catwalks, and bridges in the Orbiter Processing Facility. This is where a shuttle spends two-thirds of its total processing time before launch. The faculty's two identical hangars, which are connected by offices, labs, and storage areas, permit handling two orbiters simultaneously.

Processing the shuttle for a return to space requires 761,000 separate operations involving many of its 210,000 parts. Virtually every phase of the ground turnaround is monitored by a bank of 336 computers known as the Launch Processing System. Temperatures, pressures, flow rates, voltages, valve and switch positions—all are scrutinized millisecond to millisecond. Readings outside the norm trigger an immediate response from the computer, which alerts the nearest technician. "There are so many parameters that have to be within spec to launch the vehicle that it's—I don't want to say humanly impossible—but it would be just a hell of a chore for the system engineers to monitor all those parameters," explains Mike Leinbach, one of NASA's test directors. "Therefore it's all computerized."

But the pre-launch phase isn't a cookie-cutter process. Each mission leaves an orbiter in a different condition: brakes may be worn, tiles chipped, or oil slobbered in the payload bay from a broken hydraulic line.

When the shuttle returns to the Orbiter Processing Facility, the first task is to complete the "safing" operations begun in the desert—draining leftover propellants and purging the lines. "There are a lot of major operations that happen right away, like pulling out the engines," says Jennifer Webb, a NASA operations engineer who graduated with a bachelor's in engineering the same year that the *Challenger* blew up. She is young enough to think of Sputnik as ancient history and to consider a career in space exploration as nothing unusual.

"If we're slow getting the engines out, then we won't have as much time to work on them, which means it will be slow getting them back in," says Webb. "This can snowball." Pulling the engines for servicing is just the beginning. To prepare for its mission to launch the Hubble telescope, *Discovery* underwent some 36 modifications during its previous stay in the Orbiter Processing Facility last winter. New carbon brakes were installed to provide greater stopping power and control during landing. The high-pressure oxidizer turbopumps on the main engines were equipped for the first time with sensors to provide data on bearing wear.

Now two months later *Discovery* hangs motionless above the floor once again. Technicians routinely check for damaged items. Their tools range from a simple lug wrench to a "smart" torque wrench, which gives a digital display of the force being exerted and an audible click when the correct torque has been applied. It's just one of more than 1,700 special shop aids available.

Often tools have to be hand-crafted to repair the 24,000 tiles that cover each shuttle's underside, wings, and other areas exposed to high temperatures during reentry. Early shuttle missions were plagued by tiles they kept falling off, but new materials and advanced bonding techniques have ensured that tiles are seldom lost anymore. Still, they have been damaged by ice shaken loose from the external tank at liftoff, stones kicked up from the runway at landing, and even rain encountered during the ferrying flight back to Kennedy.

Replacing tiles is a very precise operation. Measuring and calculating the placement of each one takes about two hours. "In the construction business, if you mess something up you can go back the next day and fix it," says Greg Grantham, who built aluminum store fronts for a living before he became a shuttle tile technician. "Here they might launch it the next day."

In addition to the repair work on the orbiter's structure, its payload bay must be cleaned and reconfigured for the next flight. Payload bays sometimes return full, as was the case last January when *Columbia* brought back the Long-Duration Exposure Facility, a schoolbus-size satellite rescued from a decaying orbit. Because the bay's hinges aren't designed for use in normal gravity, special counterweights must be used to simulate the zero-G condition of space.

Technicians wearing plastic gloves and hoods covering their hair prepare the bay for new payloads. Some are loaded while the shuttle is horizontal. Those that can be installed in the vertical position will be added when the orbiter is on the launch pad.

Riding herd in the Orbital Processing Facility are the flow directors, the managers who pull the various jobs together. Each of the three shuttles has its own flow director. Their main adversaries are the "pokers," problems that "poke out of the flow, out of schedule, so you've got to work around them and accommodate them," says Tip Talone, flow director for the *Discovery* shuttle. "If we run into a problem," he explains, "we will just replan it and push the launch date back."

Work continues around the clock, three shifts a day, seven days a week, by thousands of contract employees. If they lived in Detroit, many would probably be building cars. Because they live near Cape Canaveral they work for private companies that perform much of the work for NASA. The prime contractor for launch preparation is Lockheed Space Operations, which works with teams from EG&G, Morton Thiokol, McDonnell Douglas, Rocketdyne, and United Technologies.

NASA operations engineers like Jennifer Webb grade the performance of the contractors. While she encourages and advises, she doesn't command them. That's left to her contractor counterpart. "Sometimes it will be very calm; it just works like a clock," says Webb. "Sometimes all hell lets loose. This part's wrong and that guy doesn't have the right piece. There's not enough people to do an operation and an engineer left for the day and his papers are all hunked up."

For two to three months the shuttle lies horizontal in the Orbiter Processing Facility. While it is serviced a metamorphosis has been taking place less than half a mile away in the Vehicle Assembly Building. Another crew has been stacking the two solid rocket boosters on a mobile launcher platform, where they will be mated to the 15-story-tall external tank. When it's ready, the orbiter is towed to the assembly building to take part in this extraordinary transformation.

One of the largest buildings in the world—its floor covers eight acres—the assembly building is the former home of the Apollo Saturn V rockets and the heart of Launch Complex 39. Here more than 70 lifting devices, including two 250-ton bridge cranes, will transform the prone orbiter into a vertical rocket ready for launch.

It's a secretive rite of passage. Guards are posted at the entrances as workers attach cables to the orbiter and set the winches. A crane-like hoist pulls the 87.5-ton orbiter into a vertical position until it dangles like a puppet 200 feet above the floor. Then it is positioned above the mobile launcher platform. As it is inched downward, its wings slide into place beside the external tank and two solid rocket boosters. Struts on the external tank align with a bracket near the orbiter's nose and two other brackets by its wings. These three connecting points are then fastened with bolts 28 inches long and 3.5 inches in diameter. The delicate process takes about 18 hours.

The orbiter will remain upright until launch. Now it's time to test connections between the orbiter, the external tank, the solid rocket boosters, and the mobile launcher platform. Hundreds of electrical wires run through five pipes and hoses that connect the orbiter and the external tank. This is the first time that an all-up integrated test can be run, and it lasts about five days.

The final activity in the assembly building is the aft installation of ordnance: explosive bolts that the orbiter will require when it's time to break free from the launch pad.

Launch pad 39-A, NASA's workhorse, sits beside the Atlantic Ocean four miles from the Vehicle Assembly Building. To move the shuttle from one to the other, Kennedy personnel use one of the biggest land vehicles in the world: the crawler-transporter.

Designed to lift, hold, and move the largest, tallest, and heaviest known portable structures on earth, the crawlers (NASA has two) were originally used to transport the Saturn V moon rocket. Sixteen hydraulic jacks—four at each corner—lift the 4,000-ton mobile launch platform and 2,250-ton shuttle. The crawler itself weighs 3,000 tons and is powered by two 2,750-horsepower diesel engines.

When it's loaded the crawler has a top speed of 1 mph, and the trip to the pad takes six hours on a roadway—the "crawlerway"—almost as broad as an eight-lane turnpike. Both crawlers now have hundreds of miles on odometers that first started turning in the Apollo years. Once it arrives at the launch pad, the crawler places the launch platform on a set of pedestals that position the shuttle engines and solid rocket motors above a reinforced-concrete flame trench.

Somewhere between two to four weeks before launch, the computer begins pumping fuel—monomethyl hydrazine and its oxidizer, nitrogen tetroxide—into tanks supplying the maneuvering jets on the exterior of the shuttle. In the early days of the program, these fuels were loaded by workers in white rubber SCAPES—self-contained atmosphere protection ensemble suits—who opened and closed valves by hand. "We have gradually changed the hardware and written the software until now most of that is automatic," says S.W. "Buz" Brown, manager of operations at launch pad 39-B.

"We get a lot of small leaks," says technician Morton Higgs. "If it runs on the floor, we have sponges to pick it up and put it into a bucket." Higgs uses the equivalent of a wet/dry vacuum when a thin stream of hydrazine dribbles down the fuel line. The worst leak happened almost a decade ago, when a fuel line connector broke. "We had propellant flowing down the belly of the orbiter," recalls Brown. "We had a lot of tiles to replace."

Nine hours before launch the computer starts pumping liquid oxygen and liquid hydrogen into the two compartments of the shuttle external tank. The top compartment holds 143,000 gallons of liquid oxygen. The bottom, more than two times larger, holds 385,000 gallons of liquid hydrogen. Fueling takes about three hours.

Last summer hydrogen leaks resulted in the temporary grounding of the entire shuttle fleet. *Atlantis*, which was scheduled to fly a classified defense department mission, suffered from a leak in a 17-inch-diameter pipe that carries liquid hydrogen from the external tank to the orbiter's main engines. A similar leak plagued *Columbia*, which was preparing to deploy NASA's Astro-1 X-ray and ultraviolet observatory.

Leaks aren't the only problems that occur on the launch pad. Last April one of *Discovery*'s auxiliary power units failed at T minus four minutes. The replacement APU came off the *Atlantis*—at the time in the processing facility being prepared for a July launch. Even though the practice of cannibalizing parts was soundly criticized by the presidential commission that investigated the *Challenger* tragedy, it still occurs and even proved to be the solution to *Columbia*'s hydrogen leak—NASA replaced the leaky hardware with parts from the shuttle *Endeavour*, which is currently under construction.

"You try to avoid it wherever possible, but it does happen," says Jennifer Webb. She recalls another incident when a transponder, or radio beacon, was taken off *Discovery* for the *Atlantis* flight in February 1990, even though *Discovery* was scheduled to fly in less than two months. "If we hadn't been into a countdown situation I don't think they would have done that," she says.

Columbia had a problem with cannibalization after the *Challenger* accident. "We kept taking parts off of it," Webb says. "We took an entire antenna for a *Discovery* flight. It gave up a lot of parts before it flew [in August 1989]."

The decision to cannibalize is made at the level of flow director or higher. The managers, Webb says, have to establish priorities. "Like what they say at the launch pad [when they request parts from a shuttle in the processing facility]: 'They've never launched one from the horizontal yet.' So we say, 'Okay, you can have it.'"

After months of processing, the shuttle is now ready, and the launch support team watches its television monitors for the pre-programmed checks. The flaps and wing elevons, speed brake, rudder and ailerons all wave to the controllers as the computer puts them through their paces.

In the final minutes prior to launch, computers take over. At T minus 6.6 seconds, the main engines ignite. The shuttle rocks 25.5 inches toward the external tank, then back as the main engines reach 104 percent thrust. As soon as the shuttle is exactly vertical, at T minus 0, the solid rocket motors fire. Explosive bolts holding the shuttle to the pad splinter and drop into sand pits below. Umbilicals release.

Gathering momentum, the shuttle lumbers off the pad. As the spacecraft clears the tower, Houston takes control of the mission. For those at the Kennedy Space Center it's a gratifying instant—and time to get back to the other two shuttles awaiting their turn.

MOONSTRUCK[2]

The landing of human beings on the moon on July 20, 1969, was a momentous event by Earthly standards, but it was still more so by lunar ones. Almost nothing ever happens on the moon, so the slightest incidents are immortalized. With no atmosphere to erode them away, Neil Armstrong's and Buzz Aldrin's footprints will probably remain intact for hundreds of thousands of years. In fact, a detailed chronology of everything that ever happened on the moon is inscribed on its surface—if you know how to read it.

For 360 years before the *Eagle* landed, ever since Galileo built his first spyglass, researchers had been trying to read that record. The moon was the nearest and most tantalizing other-world, fatly filling a telescope's field of view; one could not help but wonder how it got there and how it came to look the way it does. What's more, as geologists came to appreciate just how long and complex Earth's history was and how much of it had been expunged by erosion and a shifting crust, the moon became more interesting. Its rarely cleaned slate, they hoped, might yield clues not only to its own distant past, but also to Earth's.

Such were the hopes that preceded Apollo. Centuries of merely looking at the moon had brought only modest returns; now at last researchers would be able to hold moon rock in their hands, compare it with Earth rock, analyze it chemically and mineralogically, measure its age, unravel its story. July 20, 1969, marked the beginning of a revolution in moon studies.

Yet it was not quite the revolution that some had expected. "When I first went into the space program, I'd just got out of grad school," recalls planetary geologist James Head of Brown University. "This was 1968, and I was thinking 'God, we're going to the moon! and—God, they're going to land on the surface! and—

[2]Reprint of an article by Tom Waters/© *Discover* publications 1990. 10:90+, J1 '89. Reprinted with permission.

God, they're going to let astronauts go out and collect rocks!
What more could you ask?' Well, my predecessors, who'd been
working on this for ten years, started out with the expectation
that *they* were going to go, A; that, B, they were going to have a
crew of geologists there, and one was going to land over here and
another over there and a big rover was going to drive in between.
It was going to be fifteen landings or whatever—really large-scale
stuff. Then, of course, the project got hammered down and down
and down, until finally these two guys were going to drop down
on the surface and get out for three hours."

The two guys and the ten others who followed (including one
geologist, Harrison Schmitt) did manage to cart back 841 pounds
of rocks, and those rocks did answer some of the big questions
about the moon—but only some. After 20 years of hands-on lunar geology, there is much that remains to be learned about the
moon's history.

A case in point concerns the answer to the biggest lunar question of all: that of how the moon formed in the first place. According to a recent hypothesis, the moon was born in violence,
when a Mars-size piece of solar system debris—a planet that
didn't make it—slammed into the young Earth, blasting off a
cloud of rock that later self-gravitated into a spherical moon.

Computer simulations have shown that such a cataclysm could
have produced a moon with the right size and the right orbit—
something earlier theories had trouble doing. And this proposed
scenario had other charms as well. It explains why moon rocks are
depleted of volatile materials such as water and oxygen: they
boiled away during the impact. And it explains why the moon has
at best a small iron core, unlike the other planets: most of the iron
in the impactor would have sunk into Earth's core.

Neither the circumstantial evidence nor the computer simulations, however, are enough to verify the impact hypothesis. What
lunar geologists really would like is more hard data. From their
point of view, the Apollo astronauts left more than flags and footprints on the moon. They also left a lot of unfinished business.

The business of lunar geology—selenology—began in 1610,
when Galileo noted that the moon had two distinct types of terrain: light-colored, heavily cratered highlands and the darker,
more sparsely cratered plains that came to be called *maria* (*mare*,
in the singular), the Latin word for seas. But not until the middle
of this century did researchers come to agree on how the moon

had acquired the craters Galileo had discovered. By systematically comparing lunar craters with ones created by volcanoes and bombs on Earth, astrophysicist Ralph Baldwin showed that no volcano (nor any bomb we've made so far) could have mustered enough energy to excavate the larger lunar cavities, known as basins, which are hundreds of miles across. Only the impact of a large meteorite or asteroid could have done the job. And in fact nearly all lunar craters, large and small, are now thought to have been formed by impacts, with the heaviest bombardment occurring early in the moon's 4.5-billion-year history, a time when the solar system was still thick with debris.

The realization that the craters were meteorite prints made them invaluable to moon researchers. Each crater records the tick of a geologic clock: the more craters in a given region, the longer the crust there has been catching meteorites, and thus the older it is. By counting craters, one can make a rough guess of the relative ages of various regions on the moon. The first thing selenologists noted when they made such tabulations is that the maria must be younger than the highlands.

The question of how the maria formed, however, was not answered until Apollo landed—but then it was answered almost immediately. There had been various theories of what Armstrong and Aldrin might find underfoot when they stepped out onto Mare Tranquillitatis. One of the most colorful was put forward by Thomas Gold of Cornell: he warned that the maria were seas of electrostatically charged dust, and that the *Eagle* was in danger of sinking.

But Armstrong and Aldrin confirmed a more pedestrian view. Some researchers had long noted the similarity of the maria to the Columbia Plateau in Oregon and Washington. That plateau is made of basalt, a kind of rock that forms when lava flows out of a volcano and cools at the surface. Sure enough, the rocks Armstrong and Aldrin retrieved were basalts.

It is now agreed that all the maria were formed by a long series of lava flows during a period of intense volcanic eruptions that probably began around 3.9 billion years ago, after the worst of the meteoritic bombardment. The eruptions continued for at least 700 million years and may not have ended until about a billion years ago. The lunar lava, fluid as motor oil, rushed out over the ancient crust, filling some of the impact basins. Then it froze into the dark seas of basalt we see today.

The maria, however, are not unadulterated basalt. Although
the flux of meteorites began to taper off about 4 billion years ago,
it continues to this day. The maria have relatively few craters visible from Earth, but they have been pounded by a steady rain of
smaller meteorites, including many less than a millimeter
across—the kind Earth is protected from by its atmosphere.
These impacts have slowly bashed the basalt into a fine, powdery,
charcoal-gray soil strewn with larger rocks. Many of the rocks and
soil particles in the maria (as well as in the highlands) are brec-
cias—composed of fragments of older rocks smashed by a meteorite.

Moon rocks in general are a jumble. And in a sense, so is the
entire moon. A good-size impact can hurl rocks a considerable
distance, as the bright rays emanating from the craters known as
Copernicus and Tycho demonstrate: those rays, which consist of
rock ejected from the craters and strewn across the surface, extend for thousands of miles. (The enormous impact that excavat-
ed Mare Orientale may have flung material halfway around the
moon, where it collided with rock flying the other way; in this
view, the large bright smudge that lies exactly opposite Orientale
is where the colliding rock fell in a heap.) Some mare rocks contain bright fragments that are not derived from mare basalts at
all. Those fragments came from the highlands.

Even before the highlands had been sampled directly by the
later Apollo missions, Harvard geochemist John Wood had sifted
highland pebbles out of mare soil and identified the bright material as a type of igneous rock called anorthosite. The landings in
the highlands confirmed Wood's diagnosis. In fact, spectal measurements made by the orbiting command modules suggested
that anorthosite, a notably light-colored and lightweight rock,
was the dominant rock throughout the lunar highlands.

Finding anorthosite on the moon was not surprising—it's not
rare on Earth—but finding so much of it was. Unlike most igneous rocks, anorthosite is composed almost entirely of one miner-
al: plagioclase feldspar. It forms when a large body of magma
cools slowly, allowing the lightweight, light–colored plagioclase
to float toward the surface of the magma while heavier minerals
(some of them dark as well) sink toward the bottom.

The implication of the Apollo findings was astonishing but
unavoidable: Since anorthosite is spread all over the surface of
the moon, the moon must once have been covered with an ocean

of magma. What's more, some of the Apollo anorthosites had been dug up from depths of 30 miles or more by large meteorite impacts. Thus the primeval magma ocean must have been at least deep. When it cooled and solidified, not long after the moon's birth, it formed the crust that remains exposed in the highlands today.

The dark basaltic lava that covered this crust in the maria rose up, during a later period of melting, from below the anorthosite layer—that is, from the bottom of the original magma ocean. Mare basalt and highland anorthosite, in this view, are the yin and yang of lunar rock. Their unity was fractured in the primeval magma, and now they exist as separate layers.

Surprising as it was when Wood first proposed it, the magma-ocean concept is now generally accepted by geologists, and its implications extend well beyond lunar evolution. "At the time we went to the moon," says Wood, "the conventional wisdom was that the moon and planets had formed relatively cold, and only later heated up, due to radioactive decay inside the planets." But once it was admitted that the moon had formed hot, it wasn't long before the principle was extended to planets: Earth too once had a magma ocean. "Conventional wisdom has swung the other way," says Wood.

In the years since the moon landings, though, the magma ocean theory has undergone a lot of refinement—or at least complication. "After Apollo, with samples in hand, we had a very simple picture of the moon's evolution," says Carle Pieters, a geochemist at Brown University. "But things have gotten considerably more complex since then." The magma ocean, Pieters has found, can't explain all the highland rocks, because not all of them are anorthositic.

Pieters is studying areas of the moon not sampled by Apollo, using a device that ought to seem quaint by now in lunar geology: a telescope. By attaching spectrometers to a powerful telescope on Mauna Kea in Hawaii, and by keeping the same spot of the moon in veiw for hours at a time, she can do a rough analysis of the mineralogy of the moon's surface. And since craters bring material to the surface from various depths—the larger the crater, the greater the depth—observing a series of progressively larger craters is like descending into the Grand Canyon: you see a cross section of the crust.

It's in the large highland craters such as Copernicus and Tycho that Pieters and her colleagues have found their most interesting results. What they've found, in essence, is rock that has no business being in the highlands, because it contains heavier minerals—olivine and pyroxene—that should have been confined to the deepest levels of the magma ocean. Such discoveries do no disprove the theory. But they do suggest that after the anorthosite crust formed, and while it was being battered from above by meteorites, it was also suffering intrusions from below by material from the moon's interior—perhaps the same sort of stuff that volcanoes later spewed into the maria. "The complexity of the lunar crust," concludes Pieters, "is far greater than we ever imagined."

Yet there is a curious pattern to the complexity—a pattern to the distribution of darkness and heaviness on the moon. The pattern manifests itself in three ways. First, most of the intrusions of heavy minerals into the highlands are found in craters on the left (western) half of the moon's near side (the side we always see from Earth); craters on the right (eastern) half tend to contain mostly anorthosite, like the Apollo samples. Second, the near side as a whole has many maria, whereas the far side has almost none. Third, the moon's center of mass is closer to Earth than its geometric center, indicating that the near side is made of heavier stuff than the far side.

Pieters and other researchers suspect that all three patterns might somehow have been produced by the impact of an immense object early in the moon's history, while the crust was still crystallizing out of the magma ocean. The Procellarum Basin, a vast, apparently circular feature that is covered by the maria of the western near side, may be the disguised record of such an impact. But the idea is just speculation at this point. "We're data-limited now," says Pieters. "We have just enough data to know that ignorance isn't bliss."

Where will new information come from? Twenty years after the first moon landing, 65 groups of researchers are still working on the rocks brought back by Apollo, and much remains to be extracted from them. But the problem with those rocks is that they come from such a limited region of the moon—roughly speaking, the central part of the near side. "To really answer the questions we have about the moon you need a global evaluation," says Pieters. "You can't just use a few pieces to extrapolate to the whole."

If you have good global maps, though, you can do some serious extrapolating, using the Apollo rocks as Rosetta stones. That's the premise underlying NASA's modest plans, repeatedly shelved over the past decade or so, for a return to the moon. Although the entire moon has been closely photographed in visible light, mostly by the pre-Apollo lunar orbiters, only about a fifth of the surface has been covered at the gamma-ray and X-ray wavelengths that are crucial for determining the composition of the surface. The *Lunar Observer*, a small probe now under consideration for launch in the mid-1990s, would extend those maps to the rest of the moon. In so doing it might help confirm or deny the impact theories of the moon's origin and evolution.

The *Observer* could also do something else: it could identify good sites for another manned landing on the moon. All lunar geologists agree that what they really need is more rocks. Of course, the needs of geologists are not what drives the space program, but the idea of a return to the moon as a first step in the manned exploration of the solar system has undeniable romantic appeal. "Eventually, we will go back to the moon," says geochemist Jeffrey Taylor of the University of New Mexico. "The only question is when."

Taylor and Paul Spudis of the U.S. Geological Survey are ready now with an ambitious plan. They propose a permanent moon base staffed by astronauts and robots. The robots would crisscross the moon, doing fieldwork hundreds of miles from the base. Each one would be controlled by a geologist sitting back at the base, wearing a helmet that enabled him to see what the robot sees; an array of sensors on his body would transmit signals to the robot, causing it to mimic his motions. The vision may sound farfetched, but as Taylor and Spudis point out, the technology is not. And if it is ever realized, lunar geologists will enjoy, if only vicariously, the full-scale field expedition they expected from Apollo, but never got.

THE ROBOT'S REACH[3]

Most of the companies that supply NASA with gear for space exploration are located in anonymous modern buildings along suburban superhighways. Honeybee Robotics is an exception. Its offices are in New York's Little Italy, in the second- and third-story lofts of a brick building right next door to the Connecticut Muffin Company. A couple of blocks away is a transient hotel; go the same distance in another direction and you'll find a cluster of cafés and art galleries.

An unusual setting for this type of business, perhaps, but Honeybee has always charted a course different from that of its competitors. "Right from the start," says Steve Gorevan, the company's 35-year-old president—and a former piano tuner—"we said we're definitely not going to travel in the mainstream." In the world of robotics the mainstream means building robots to man automotive assembly lines; Honeybee opted to design robots to work in aerospace and in lighter industries like electronics and pharmaceuticals. "We also thought that, being in New York, we could do some work in the movie and television business—robotic cameras, special effects—but that didn't lead anywhere," says Gorevan.

Honeybee—named in recognition of the insect's superb engineering skills—got its first job seven years ago, designing a robot to handle the delicate job of removing freshly made contact lenses from their molds. Since then, work for NASA has gradually taken over, and Honeybee now spends 80 percent of its time on one highly prestigious project. With Ford Aerospace, it is designing the end effectors—the hands—of the most sophisticated space robot ever planned.

The robot is known as the Flight Telerobotic Servicer, or FTS, and it will assist astronauts in building and maintaining the space station. It will be able to perform such tasks as removing and replacing struts from the station's scaffoldlike truss and connecting and disconnecting hoses or electric cables. As now envisioned, the FTS will work under the direction of a human with

[3]Article by Tom Waters/© *Discover* publications 1990.11:68+, 0 '90. Reprinted with permission.

a joysticklike control, but ultimately it could work autonomously. It will be far more complicated than any previous robot, able to move as many as 26 different joints; the space shuttle's mechanical arm, in contrast, can move only six.

The FTS will also look more like what we picture as a robot: it's anthropomorphic, with a head, arms, and torso, although it has only one leg and one foot. Its foot will attach to holds on the station's exterior that will provide the robot with power and carry instructions from human or computer controllers. The robot will also be able to attach itself to unpowered holds, getting its energy from an internal battery and its instructions from an astronaut with a radio remote-control device. In the torso will be more computers and holsters for the tools the robot will use. The head will contain video cameras and work lights.

"It's science fiction come to life," says Gorevan. "The public is really going to love it."

Honeybee's involvement with NASA started with a project undertaken by chief engineer Chuck Hoberman—a 34-year-old former kinetic sculptor. At the time he wasn't trying to drum up business; he was simply amusing himself. "I had been developing a series of inventions," he says, "a way of making structures that pack down to a small scale, then unfold to be really large. But when Steve saw them, he said that we had to show them to NASA."

In 1986 Hoberman took a briefcase containing six small models and some drawings to Martin Mikualis of NASA's Langley Research Center. Unlike the Jet Propulsion Laboratory or the Johnson Space Flight Center, Langley concentrates not only on here-and-now hardware but also on theoretical projects that will be used only in the distant future, if at all.

"Before we went down there," Hoberman recalls, "Mikualis said on the phone, 'A lot of crackpots come through here and most of what I look at doesn't meet our needs.' When I got there, I just took out my models and folded and unfolded them. He didn't say anything for forty minutes. Then he said, 'It looks like origami with thickness. Maybe we can do some business.'"

The comparison with the ancient Japanese art form was apt. "I have always been fascinated with mechanisms from an aesthetic point of veiw," says Hoberman. Moreover, folding schemes were at the heart of Hoberman's elaborate paper and cardboard models. The simplest one started out as a pentagon the size of a deck of cards. As a flap at the top was pulled up, the deck extended into a tube. Then, unexpectedly, the tube turned inside out, and a second tube appeared, pointing at right angles to the original one. The whole unfolded assembly was a foot square.

Hoberman also brought along models that folded out into domes and tentlike cones. The principle behind all of them was similar to the principle behind extendable tongs, which use hinged, criss-crossing bars to transfer motion so that pushing the handles together makes the other end move forward, while pulling the handles apart makes the end move back. In Hoberman's designs, however, the structure moved in three directions. "If you pleat a sheet of paper so that there are folds running in more than one direction, you get the idea," he says. "The sheet takes on a wrinkled, three-dimensional corrugated pattern. As you press down on it, it spreads out in all directions."

With Mikualis's urging, NASA gave Honeybee a modest contract to develop a small, quickly deployable shelter for potential use at the space station. Then came a series of other projects based on space origami.

"We looked at a large hangar that would fit into the shuttle and then unfold to be larger than the entire ship," says Hoberman. These structures could be prefitted with tools on Earth and provide astronauts with ready-made workshops that would be less disorienting than the void of space. "And we looked at a reflector that might be used for a telescope or for a solar collector. We call it the starburst reflector. It unfolds to a disk with a diameter greater than sixty-five feet, but it fits in the shuttle cargo bay, which is only a little more than fourteen feet." None of these devices has yet been built, but Honeybee has completed the rough blueprints should NASA ever decide to go into production.

Honeybee also began working on more conventional robotics projects for Langley, including the design for a robot that crawls on two tracks in a Langley work area while assembling a tetrahedral truss, all under computer control. The project helped test a machine's ability to assemble an entire structure autonomously. "This was probably the first time a robot has completely built a structure from start to finish, without the help of a human with a joystick," says Gorevan.

The Langley projects gave the Honeybee engineers the chance to experiment and be creative, but what they most wanted to do was design a real, feasible space robot. So when NASA announced its plans for the space station and its FTS, Honeybee jumped at the chance to get involved. "The Langley work that we had done was right up the FTS's alley," says Gorevan. "At the time, we had experience that other people didn't have."

NASA and its prime contractor for the FTS, Martin Marietta, apparently agreed. After awarding the job of building the robot's hands and base to Ford Aerospace, they designated Honeybee as subcontractor. The shared Ford-Honeybee contract alone is worth between $10 million and $15 million.

The requirements for the robot hands were necessarily strict: they had to allow the FTS to perform a specified list of tasks, and they had to be detachable at the wrist and replaceable with new end effectors that might later be designed for specialized tasks. Also, NASA required that the end effectors be equipped with two back-up parts for every key part to ensure that the robot never lets go of anything. It's a little like equipping a vise with two safety clamps, so that if the main screw fails, the first clamp will hold it closed; and if the first clamp fails, the second one will do the job. "One of NASA's biggest worries is a part floating around loose up there," says Gorevan. "That would cause all kinds of havoc."

Nonetheless, competing designers did have some room to maneuver. NASA originally envisioned a robot with completely different right and left hands, one to hold an object and the other to work on it. "You see that the left hand has two fingers that allow it to grip," says Gorevan, pointing to a drawing posted on the studio walls, "and the right hand is a rotary device that can be made to interface with a screwdriver or similar tool."

The Ford-Honeybee idea was to allow both end effectors to have both functions. "Each hand will have a gripper drive as well as a motor that provides rotary capability," says Gorevan. "It's a more efficient way to do things. We eliminate a lot of the tool exchanges, which are risky. And we think we can reduce operation time too."

It's also a more complicated way to do things, and engineers from the two companies have spent the past 18 months perfecting their designs and building their prototype. Although the two groups work in their own shops, they communicate by fax and visit frequently. "We go over the design again and again," says Gorevan, "and it's getting finer and finer."

The robot, of course, will be worthless if its hands cannot operate reliably in the harsh environment of space. The space station will orbit Earth 16 times a day, passing from sunlight to shadow each time. As it does so the temperature of the materials will fluctuate from -40 to 185 degrees, and they will expand and contract with each cycle. The engineers need to know exactly how the hands will respond to these conditions.

"We can't actually test these wide thermal conditions on site," says Gorevan. "We can heat the parts up here and test them, but we can't cool them down—Ford's going to do that. But where we can, we simulate them. We can predict how much the metal's going to expand, so we put in a bigger piece of metal, or use screws to simulate increased expansion."

Honeybee's engineers also use computers to simulate conditions in space, but they don't rely on them exclusively. "I think you can get seduced by most advanced equipment," says Hoberman. "You lose touch with what you're really trying to understand." So at Honeybee they do a lot of hands-on work, building model after model of every part. Indeed, the engineers themselves do much of the machining.

"This way," says Steve Glappa, a 25-year-old Honeybee engineer who spends much of his time downstairs in the machine shop, "you're always bringing the design process into the building process and the building process into the design process."

So far, the most difficult component to design has proved to be the torque transducer, a small wagon-wheel-shaped piece of metal that senses the amount of force the hand is applying. When the hand grips, the wheel's spokes bend slightly—too slightly to see but enough for electronic sensors to measure. However, changing temperatures can also make the spokes bend, and these changes drown out the correct readings.

To solve the problem, the Honeybee engineers tried several different types of metal—such as stainless steel and an iron-nickel alloy—with different stiffnesses and expansion rates. They also tried several design tricks, including cutting slots in the rim of the wheel to relieve the thermal stress.

"This started out as a sophomore-level engineerng problem," says Glappa. "But it took us about six months before we really understood what was going on."

"We had to make lots and lots of modifications," says Gorevan. "You can make all the models you want on a computer, but then you have to corroborate the computer model with a real, physical model. We sat around and argued for months— 'Suppose we cut a groove here or change the shape there'—then we'd go build a model and test it. In the end we made maybe twenty models before we got one that worked"—one that, contrary to expectations, eliminated the slots and used ordinary steel.

The Ford and Honeybee engineers have now completed the design of the gripping portion of the hand. NASA will soon be subjecting it to ground testing and sometime in 1991 will fly a prototype, attached to a robot arm in the shuttle's cargo bay, to see how it does in the weightlessness of space. The gripper will be put through its paces attaching and detaching shortened versions of the space station's struts and hubs. At Honeybee the race is also on to be ready for a second test flight, in 1992, when a complete hand, including the rotary capability and the wrist detachment system, is scheduled to fly. After that the hand is supposed to be ready for use on the space station—whenever the space station is ready, that is.

The success of the Honeybee engineers in the space field hasn't diminished their enthusiasm for more mundane robot applications, as evidenced by Robotender, the robot mixologist. Gorevan believes the partially autonomous bartending robot is the perfect first step in designing more elaborate robots that could keep house, care for lawns, or pick up the garbage on a city street.

"You can't just put a robot in a home or garden and expect it to run around working," says Gorevan. "The environment is too unstructured. But a bar is incredibly structured: the bottles are lined up, the soda machines, ice machine, wine and beer taps—it's all very tightly controlled."

Moreover, compared with other robots, Robotender is cheap to build. Industrial robots almost always come with customized mechanisms, tools that represent a large part of the robot's cost. Robotender is designed to work with the same simple tools human bartenders use. Its robotic arm is also off-the-shelf, a standard industrial model that would be equally at home making cars or toasters.

In its current version, Robotender sits in the middle of a circular bar, so that it can reach its supplies by simply pivoting and extending its single hand. However, Honeybee designers do have sketches for a robot that could run on a track up and down a more conventional bar.

Robotender is equipped with a custom hand, pressure and weight sensors, and software that enables a bar owner to decide what drinks the robot should make and how it should prepare them. When a waiter types in an order, the program looks up the step-by-step procedure for making that drink and choreographs the arm's movements.

Those movements can be disconcerting, an odd mix of stunning speed and meticulous restraint. If someone orders a martini, for example, the arm abruptly swings over to a rack of martini glasses, slows, picks one up gingerly, then zooms back to the bar and sets it down. It continues to alternate fast and slow motions as it lunges for a martini shaker, picks that up, brings it to the dispensers that release measured portions of liquor from bottles held upside down, and adds vermouth and gin.

The next trip is to the ice machine; Robotender can tell whether the correct amount of ice has been added by the weight of the mixture in its hand. Once the ice is in, the robot stirs the mixture, then carefully taps the stirrer on the rim of the martini shaker to knock off the last drops. (James Bond would be glad to know that Honeybee plans to add the choice of shaking the drink instead of stirring.) Finally, Robotender gets a screen to cover the top of the container and pours the drink into the glass.

Robotender can do more than just mix drinks. The last and perhaps not least important of its components is a television set that serves as the robot's head and gives it a personality. The screen shows a bow-tie-wearing human head and neck. Robotender introduces itself, asks the customers what they would like, tells jokes, and advises against drinking and driving.

"It's really up to the bar owners what they want the head to do," Gorevan says. "If it's a sports bar, they might just want to put the game on. We can show videos, whatever." Indeed, Robotender has even been programmed to dance while showing Robert Palmer's "Addicted to Love" video.

"At first, I was a purist about this," Gorevan says. "I didn't want the television on its head. I just wanted it to work." However, another objective was to get the robot into an actual bar where people could see it work. Gorevan hopes that bar owners will want to invest in the robot as a form of entertainment.

The next step, says Gorevan, could be a robot designed for a diner or a dry cleaner—"we have to study the situation."

Of course, developing robot bartenders and short-order chefs is very time-consuming, and the Honeybee engineers are keeping these projects on the back burner. At the moment, they have more important matters to consider.

"Since we got the contract for the FTS, we've been going constantly, trying to get ready for the first test flight," says Gorevan. "Now we have to work even harder to get ready for the second flight. But this work is going to have a lot of payoffs: Private industry is very interested in FTS because the robots are so innovative. Manufacturers could use FTS-like spin-offs. The nuclear industry could use them to clean up areas of their plants that are too radioactively hot for people to enter. FTS could be with us for a long time—even after the space station is built."

THE NEW SPACE RACE[4]

On the evening of April 7, peasants clustered on the darkening hillsides surrounding China's Xichang launch site, waiting. Towering more than 43 meters above the launch pad sat China's Long March 3 rocket, ready for lift-off. Its mission: to boost *Asiasat I*, a communications satellite made by Hughes Space and Communications Group, into orbit.

The spectators did not wait long. Indeed, the launch may have been the easiest part of the exercise. Chinese officials had already weathered a political tempest in Washington, D.C., and secured special permission from President Bush to launch the American-made satellite. At 9:30 P.M. Long March soared eastward, jettisoning its payload 21 minutes, 27 seconds later. Within weeks, the satellite began transmitting data, television broadcasts and telephone conversations from countries across East Asia. Equally important: the launch put China among the runners in the space marathon of the 1990's.

The first space race was a contest to prove technological prowess and military might. Now the missiles of the past have

[4]Article by Elizabeth Corcoran and Tim Beardsley. *Scientific American*. 263:72+. Jl '90. Reprinted with permission. Copyright 1990 by Scientific American, Inc. All rights reserved.

blossomed into commercial launch vehicles. Even though national prestige remains important in the ongoing space race, this time the runners are struggling for positions that will command future commercial opportunities. Some half a dozen nations are competing to build businesses—or at least trying to earn some money—by launching commercial payloads. But they face a big problem: "commercial space" is still an oxymoron.

There are too few customers to fill all the open slots on launch schedules. Communications satellites are now the only space venture that the private sector can afford, but the world's appetite for these will soon be sated by about 15 launches a year, according to the Paris-based research company Euroconsult. In contrast, the number of available rides to space this year is closer to 40. Making matters worse, satellites are lasting longer because their components are becoming more durable. Fiber-optic cables are an attractive earth-based alternative for long-distance transmissions (except in remote areas such as Indonesia).

Even so, governments are determined to stay in the space race. Communications and satellite intelligence have become critical not just to national security but to countries' economic health. And then there is national pride. "We have a choice now," says D. Allan Bromley, President Bush's science adviser and a member of the National Space Council. "We can either be a major player in this new frontier, or we can pull back."

So nations are turning to the commercial arena to subsidize their launch capability. New—albeit uncertain—markets hold promise. Among the possibilities are direct-broadcast satellites for television and satellites for consumer navigation systems. The military is exploring "lightsats," small satellites weighing a few hundred kilograms that could replace systems damaged during a crisis. Networks of these could also be used in civilian communications.

And then there is the elusive goal of manufacturing in space. The best hope rides with experiments that investigators and small companies are now putting into orbit. Given a decade or so and enough experiments, some might prove that working in a microgravity environment is lucrative.

Outside the U.S., governments are working with industry to finance space programs. European countries, particularly West Germany and Italy, are vigorously pursuing microgravity experiments. So is Japan. The Soviet Union—with its fleet of reliable rockets, its *Mir* space station and its reentry capsules for bringing experiments back from space—seems to have the most robust space infrastructure.

In the U.S. the world's largest rocket makers await signals from the government. Large corporations are shying away from experimenting in space because their calculations indicate the risks are too great. As a result, start-up companies seeking niches in space look promising. And they are learning a lesson the large companies have yet to take to heart: building a launching business means developing a service, not just hardware. "You can't just build metal, add 10 percent to its cost and call it a business," says Charles Chafer, who in 1980 helped found Space Services in Houston, Tex.

NASA Fumbles

Unfortunately, slapping a price on a piece of metal has been the way the space business was run for decades. Through the 1970's any company in the world hoping to launch a commercial payload had one choice: the U.S. National Aeronautics and Space Administration. The U.S. offered variations of a quartet of vehicles: heavy-lift Titans (built by Martin Marietta), midsize Deltas (from McDonnell Douglas) and Atlases (from General Dynamics), and tiny Scouts (made by LTV Aerospace and Defense Company). The contractors built the vehicles; NASA acted as a broker and ran the launches.

Then NASA tried to replace these expendable vehicles with the space shuttle. It had argued the shuttle would be a cheaper way to get into orbit. (This proved true only when the launches were subsidized; one estimate of the cost of a dedicated shuttle flight today is far more than $350 million.) As a result, commercial and almost all military satellites were booked onto the shuttle. But NASA had trouble running a commercial launching business. Flights were often delayed, upsetting the plans of satellite owners.

So many satellite builders began looking to the fledgling European rocket Ariane 1 as an alternative. Incorporated in 1980, Arianespace's founders took a businesslike approach from the beginning. They deliberately kept the bureaucrats of the European Space Agency (ESA) out of the day-to-day business of building and launching rockets. Arianespace helped customers arrange financing for flights. By the end of 1985 Arianespace had won about half the commercial contracts for launching satellites. "NASA treated its clients like a lord accepting peasants on his land," asserts Charles Bigot, managing director of Arianespace. "Arianespace treated them like a retailer."

Although the Reagan administration permitted U.S. rocket makers to compete for commercial launching contracts, not one was willing to take on the shuttle. The explosion of the *Challenger* shuttle in 1986 changed that; the U.S. government subsequently agreed to ban NASA from launching commercial payloads. But U.S. contractors still faced stiff competition. Arianespace had won a loyal following and was steadily improving its vehicles. ESA covered the heavy technology-development costs. Arianespace earned its operating expenses with satellite flights (and now needs to win about six or seven annual launch contracts).

Time is not helping U.S. launching companies' competitiveness. The government continues to have trouble figuring out what it means by a "commercial" launching industry. Contradictory priorities from different federal agencies buffet U. S. policy. The military, anxious to maintain a mixed fleet of rockets, has awarded almost enough contracts to sustain the three large-booster companies for the next five years. But the contracts may have dampened the companies' interest in ferreting out commercial contracts.

Mixed Signals

"Most of our customers are repeat customers" who flew payloads on board Deltas when NASA ran the launches, says Samuel K. Mihara, marketing director for McDonnell Douglas's Delta. One commercial launch a year would be average business he says. More than one would be good. Martin Marietta has secured 42 firm contracts for Titan launches, scheduled to take place through the late 1990's. Forty-one are devoted to Air Force cargo; one will carry NASA's *Mars Observer* probe. Industry sources suggest that Marietta may drop out of the commercial the market altogether—a rumor the company denies.

The mixed signals have not helped the traditional aerospace contractors adjust to their new roles. "To this day, I still haven't figured out what the hell commercial space is," says Jack Whitelaw, a manager for LTV.

Nor have the large U.S. companies been willing to lend much support to start-up companies keen to venture into space. "Large U.S. aerospace firms have not played a strong strategic investment role in new commercial space ventures," notes a recent report from the Department of Commerce.

So even small U.S. companies planning space ventures are looking overseas for funds. SPACEHAB, a company in Washington, D.C, that is building a module to fly on the shuttle, found willing investors in Japan. Geostar, also in Washington, D.C., which is developing a satellite-based tracking system for the trucking and railroad industry, similarly relies on support from Japanese and French investors. Meanwhile other governments have begun hoping that commercial launches may supplement their space budgets or add to their hard-currency earnings. Both the U.S.S.R. and China are now wooing commercial satellite customers with the hallmark tactics of capitalism—low prices and polite service. These countries worry more about the political barriers to winning contracts than breaking even on launches.

Initially the Soviets tried to attract commercial satellites by offering bargain-basement prices. They were routinely rejected. Now Glavkosmos, the Soviet marketing arm for space services, has honed its pitch. The Soviets are betting that the reliability of their rockets and diversity of space infrastructure will draw customers. Glavkosmos has had some success. INTOSPACE in Hanover, West Germany, has flown experiments with the Soviets, as has Payload Systems in Cambridge, Mass.

The Chinese, on the other hand, are still apparently offering cut-rate prices—which would violate their pledge to the U.S. to sell services at prices comparable to those charged by the West. A recent Chinese bid to fly an Arabsat communications satellite "was about half of what we or McDonnell Douglas would have to charge," asserts Douglas Heydon, president of Arianespace's U.S. subsidiary. The office of commercial space in the Department of Transportation was investigating the concerns in early May.

And yet another competitor, namely, Japan, is coming up. Its National Space Development Agency (NASDA) has managed a small but rapidly growing launch program since 1969. In exchange for U.S. rocket technology in the 1970's, Japan agreed to launch only Japanese payloads. But a new rocket, the H-II, which is set for its first launch in 1993, is based entirely on Japanese design. This will free Japan to launch other nations' satellites.

Although Japanese officials maintain that their interest in promoting a commercial H-II program is slight, 13 H-II contractors, led by Mitsubishi, are due to establish formally a new company late this month to develop the rocket for NASDA. "Japan began development of the H-II with the knowledge that commercial launches would occur in the future," says Koji Sato, a member of Mitsubishi's coordination office for the new company.

Hurdles remain. Fisherman working near the Tanegashima Space Center convinced the government to limit rocket launches to two 45-day periods every year, so Japan is watching with interest the development of a commercial launching center in Australia. Moreover, the appreciation of the yen has helped push the estimated launching costs for the H-II out of the commercial arena. "If cost reductions cannot be achieved, this company will remain at a low profile, just selling hardware to the government," Sato says.

Marginally reducing the cost of flying will not change the private sector's disinterest in working in space in the near term, however. It will simply help current users reduce their initial capital expenditure. "Without a valid business purpose, no one will go to space at any price," says Albert D. Wheelon, retired chief executive of Hughes Aircraft. Expensive rockets are bought only by governments and commercial satellite companies. The U.S. government spent about $25 billion on space goods and services in 1988. That same year the commercial sector spent about $1.8 billion (and $2.7 billion in 1989), more than 90 percent of which is attributable to satellite communications.

From a corporate perspective, most space ventures must return revenues that are at least double their costs because of the risky nature of the business. Last February, for example, a failed Ariane 4 launch deposited two Japanese satellites worth $200 million into the ocean. In March, a commercial Titan put a $150-million *Intelsat 6* satellite into a useless orbit. Consider space from the vantage of GTE Spacenet Corporation in McLean, Va. To replace three of its aging satellites, GTE has budgeted almost $200 million to build the spacecraft, at least $210 million to launch them and another $70 million for insurance. That puts GTE's total cost for one mission at roughly $160 million. On the other side of the ledger, however, GTE expects the satellites to recover their costs in less than three years and to continue generating earnings for another seven.

Satellite owners would doubtless switch their business to cheaper but equally reliable flights. "If you're chairman of an American satellite company and your choice is $80 million with Great Wall [China's national launching company], you put up the $15 million and travel to China," says Joseph P. Allen, president of Space Industries International in Webster, Tex., and a former astronaut. "It's a no-brainer."

Profit Squeeze

Even so, prices are far from low enough to turn other space activities into lucrative ventures. Remote sensing, for example, "is a good way to lose money, hand over fist," says John E. Pike, a space-policy analyst at the Federation of American Scientists. Neither the U.S. Landsat program nor the French SPOT Image is a self-supporting business. The largest planned remote-sensing program over the next few years will be funded by NASA.

Although corporate interest in manufacturing in space flared in the early 1980's, it was doused by the irregularity of launches and by new—and far less costly—earth-bound techniques for accomplishing the same task. "There was a lot of hype," recollects Earl L. Cook, who directs the 3M Corporation's space research. 3M remains one of the few large U.S. companies that still sends experiments into orbit on board the space shuttle. And NASA does not charge for these flights.

"It's almost too bad that communications satellites succeeded so fast," says H. Guyford Stever, science adviser to presidents Nixon and Ford. "It gave us a false impression. It got people thinking there must be a lot of nuggets up there," he says. They may be up there, he adds; business has not yet learned how to squeeze from them enough revenues to justify going into space.

So national space goals will continue to drive advances in rocket technology. During the next few decades a range of mammoth government projects—free-flying orbiting platforms (which astronauts would visit occasionally), manned space stations and possible missions to the moon and to Mars—will require hauling huge amounts of materials into low earth orbit.

As a result, some space-vehicle organizations are straining to increase the payload capacity and cut the costs of launching their next generation of vehicles. ESA's development of Arianespace's next family of vehicles, the Ariane 5, may be the most aggressive project. Slated for its first launch in 1995, Ariane 5 will have two stages and rely on a main engine fueled with liquid hydrogen and liquid oxygen, supplemented by solid boosters. This should enable Ariane 5 to ship 1.5 times the payload weight handled by its predecessor into geotransfer orbit. Ariane 5 will have an option for its second stage: replacing the unmanned cargo bay with the *Hermes* manned spaceplane.

Although Arianespace relies heavily on proved technologies, innovations are working their way into Ariane 5. The solid-booster nozzles are lined with carbon-carbon, an advanced composite of woven carbon fibers impregnated with resins and then sintered. Ongoing tests are evaluating a carbon-silicon carbide nozzle for the liquid engine. To lower the operating costs of the larger vehicle, Arianespace is planning a mobile, ground-control system that will trim the time between launches and reduce the crew needed. As a result, Ariane 5's operating costs may be only 85 percent of those of Ariane 4—about $12,000 to $15,000 per kilogram to geotransfer orbit and half that to low earth orbit, says Arianespace's Heydon.

Japan's H-II will employ a complex liquid hydrogen-oxygen system in an engine called the LE-7. The LE-7 "is very efficient, but its development is very difficult," concedes Masafumi Miyazawa, who directs NASDA's propulsion systems group. Two disastrous tests of the engine last year pushed back the H-II development schedule some 12 months. Still, Japan has already successfully built a highly reliable second-stage engine—the LE-5A—that can reignite in orbit. The cost of developing the H-II will be more than Y 250 billion ($1.7 billion), $470 million of which is devoted to the LE-7.

The future U.S. launching picture looks bleak in comparison. Because of the policy of relying on the shuttle, the U.S., although long the world's leader in rocket innovations, spent virtually nothing on other advanced propulsion and launch systems between 1972 and 1986, points out John M. Logsdon of George Washington University. The Aerospace Industries Association is seeking redress by proposing that the government spend $5.5 billion over 10 years on rocket research—and is invoking the specter of foreign domination of the launch market to buttress its demands.

Current U.S. aerospace materials research aims to satisfy the demands of the National Aero-Space Plane (NASP), an air-breathing vehicle that would take off and land like a plane but travel at an unprecedented 25 times the speed of sound and go into orbit. A prototype is unlikely to be built before the year 2000. Still, NASA hopes that high-performance materials being developed for the plane will benefit rocket engines. These include copper reinforced with fibers of graphite, tungsten or niobium for conducting heat. Silicon carbide, silicon nitride and aluminum oxide fibers are being used to strengthen ceramics. Even more heat-resistant ceramics, based on zirconium and hafnium, are in the works.

Until recently the Advanced Launch System (ALS) program, initially proposed for the Strategic Defense Initiative Organization, was the principal U.S. rocket-technology program. In theory, the ALS would be a family of reliable rockets to boost as much as 200,000 kilograms into low earth orbit for as little as $660 per kilogram. The effort, originally expected to go on through the 1990's and cost up to $12 billion, was soon derided by critics as too ambitious, too expensive and unnecessary. Last December, under pressure from Congress to cut its budget, the Air Force dropped the effort.

New Elements, Old Vehicles

The ALS project aimed to trim launch costs by designing components for easy manufacturing and by employing more automation for testing parts. Martin Marietta is developing new liquid hydrogen and oxygen tanks made from an aluminum-lithium alloy that could be extruded or cast instead of milled. These might be used on future Titans. Rocketdyne has designed a fuel pump with only one weld; comparable pumps on the shuttle have 150.

The government also hoped the ALS contractors would incorporate such innovations into their current vehicles. To some extent, they are. Martin Marietta says the ALS technologies will help reduce Titan IV launch costs by one third to about $4,400 per kilogram to low earth orbit. Engineers note, however, that only limited efficiencies can be gained by trying to fit new elements into old vehicles.

Although the full ALS agenda has been canceled, NASA and the defense department are still funding rocket manufacturers Aerojet and Rocketdyne to develop a prototype engine to the tune of $107 million in fiscal 1990. The engine "is supposed to be a cheaper, expendable liquid oxygen and hydrogen engine," says James R. Thompson, Jr., deputy administrator at NASA. But because the work is funded annually by Congress, its future is uncertain.

In lieu of a new vehicle, NASA is half-heartedly considering refitting the liquid-fuel engines and solid-fuel rocket boosters from the shuttle onto an unmanned cargo vehicle dubbed *Shuttle C.* The program does not hold many technology advances for the U. S., however. Even these tentative plans are in limbo. "Frankly, it will all come back into focus once we all agree—and this is going to take several years—on a schedule to implement the space exploration initiative"—the manned missions to the moon and to Mars, Thompson says.

While NASA sets its sights on Mars, tiny U.S. companies are betting they can earn a living by putting payloads of a few hundred kilograms into low earth orbit. But first they have to improve their track record. The earliest space start-up, Space Services, conducted a suborbital test flight of its Conestoga solid-fuel rocket in 1982 but has yet to run a successful orbital flight. An attempt by the American Rocket Company in Camarillo, Calif., to fly an innovative hydrid-fuel vehicle late last year ended in a smoldering heap. E'Prime Aerospace Corporation in Titusville, Fla., hopes to piece together sections of Peacekeeper missiles.

At the front of the pack is Orbital Sciences Corporation in Fairfax, Va., with its Pegasus vehicle. Pegasus flaunts a triangular wing and an unusual launch technique: it rides beneath the wing of a large aircraft such as a B-52 or 747 to about 12,000 meters and takes off horizontally from there. In April, Pegasus lofted its first cargo into a polar orbit about 600 kilometers high—two experiments weighing 220 kilograms in all.

Orbital Sciences adopted the strategy that has worked well for the major rocket companies; it has secured the U.S. government as an anchor client. The Defense Advanced Research Projects Agency largely sponsored that first flight and is also helping support Orbital Sciences' Taurus, scheduled for flight next summer. This vehicle, which will take off with only 72 hours of preparation, will eventually loft more than 1,360 kilograms into low earth orbit.

For now, the start-ups are pinning their hopes on lightsats, a new class of small satellites that weigh only a few hundred kilograms. From the vantage of the Defense Department, which is funding much of this work, lightsats could quickly replace satellites destroyed or jammed during a military crisis. Lightsats will augment large military satellites, says Terry A. Higbee, program manager for lightsats at the Ball Aerospace Systems Group, who predicts demand in the U.S. could reach 10 per year by the mid- to late 1990's. The small launch companies hope there will be a commerical market as well. Last February, Orbital Sciences requested permisson from the Federal Communications Commission to construct its own constellation of 20 lightsats for two-way commercial alphanumeric communications.

Eventually those companies may be positioned to capitalize on renewed interest in microgravity. "Some day there'll be a lot of commercial activity in space," says Laurence J. Adams, retired president of Martin Marietta. "But we need to do a lot of experimentation with all the ideas that people have."

Europe seems to be trying to do just that. ESA is devoting more of its money to microgravity experiments than NASA is: in 1988 the European agency spent $98 million, or 2.6 percent of its budget, on microgravity. By comparison, NASA spent $63 million (.7 percent). Japanese, West German and Italian companies are also investigating the effects of microgravity on materials.

Since the advent of the space shuttle, the U.S. has offered limited opportunities for prolonged experimentation in microgravity. One reason: NASA mothballed its 20-year program in returnable reentry vehicles (space capsules) and only recently began backing this technology again. Western researchers have made some use of the available alternatives, namely, returnable Chinese and soviet capsules. Pay—load Systems has also flown materials experiments on board the *Mir* space station. But trade restrictions still limit these options.

Small European and American companies and university researchers are also trying to revive returnable capsule technology. To foster longer-term experiments, ESA is building a free-floating platform called *Eureca* and an autonomous laboratory dubbed *Columbus* that will be serviced by the *Hermes* spaceplane every six months. The purpose: to "validate a large quantity of result in order to persuade industry to commit itself to operational uses," according to Euroconsult. (Two other *Columbus* modules are in the works as well: a unit for the space station *Freedom* and an orbiting polar platform.

In contrast, NASA is placing its bets on *Freedom*. Private industry schemes such as the *Leasecraft*, proposed by the Fairchild Space Company, or the *Industrial Space Facility*, promoted by Space Industries, have been stranded in a political and financial no-man's-land. Any proposed platform would need a well-heeled anchor tenant—most likely the government—to help recover costs and to convince other investors that the project is real. But because supporting a research platform could weaken NASA's argument favoring a space station, the agency has been skittish about funding platforms.

Space is no longer a match between the U.S. and the Soviet Union. Neither has it become a commercial free-for-all. That will change only when entrepreneurs and researchers demonstrate the commercial potential of space.

They cannot do so on their own, however. Governments must reduce the risks borne by space-bound entrepreneurs—and can do so by supporting research and building accessible infrastructure in space. It is a challenge that has some urgency; as the force of military necessity wanes, governments must prove their interest in commercial space goes beyond rhetoric. Those nations that succeed will see their team pull ahead. Those that do not will be left in the dust.

HIS HEAD IN THE STARS[5]

For all our exploring, space is still a mystery; one tiny, pock-marked moon, after all, is as far as we've set foot. Perhaps that is why we demand more than just dry data from those assigned to study the planets and peer beyond the Milky Way. A space scientist has to be a visionary, a poet in a white lab coat who can give voice to our collective craving for adventure, our fascination with a universe we have not been able to touch.

[5]Article by Michael Norman. Copyright © 1990 New York Times Company, p44+, My 20 '90. Reprinted with permission.

At first sight, Ed Stone is not such a man. As he hurries through the luminous California morning, one hardly notices him, 5 feet 10 inches, 130 pounds, a wisp in gray—gray suit, gray shirt, gray felt shoes—lugging an ancient leather briefcase. He keeps to the shadows and side paths, terra incognita, a physicist so swept up in his daily occasions, so occupied by science, his life appears to turn on little else.

And yet, for the public, Ed Stone has been a kind of Cicero on space. As chief scientist on Project Voyager, he has participated in some 60 public briefings and news conferences over the 13 years of the project. Since 1977, when the two Voyager spacecraft began their parallel mission to explore the far reaches of the solar system, Stone and a team of fellow scientists at the Jet Propulsion Laboratory, the country's leading center for planetary exploration, have nudged the small spacecraft from one outer planet to another—past Jupiter, Saturn, Uranus, Neptune—extending our presence and, through stunning photographs, our line of sight into the Milky Way.

Modern science is usually big science, with apparatus so large and complex it takes teams of scholars and technical experts to make it work. Voyager at one point employed as many as 120 scientists. Ed Stone's leadership on that project has served as a paradigm for such big-science undertakings. With Voyager, Stone was the first director of spacecraft science to bring extensive scientific expertise to management. In doing so, he "revolutionized the world of project science," in the opinion of his boss, Norman Haynes, who for three years was Voyager's overall project director.

Now Voyager is traveling out beyond the solar system. In the last stage of its long mission, the spacecraft is headed for the heliopause—that point in the cosmos where the sun's influence ends—frontier to the great tenebrous firmament of intersteller space. As it travels away from the sun and beyond the planets, Voyager will send back data for the next 25 to 30 years, at which point its fuel—hydrazine for maneuvering power and plutonium for heat and electricity—will be spent. Its mission now is to reach the edge of the solar system and dip its probes into deep space.

If its power holds and its systems survive, Voyager is expected eventually to make the frontier, perhaps in 20 years.

Last winter, scientists turned on Voyager's camera for the last time and commanded it to capture one final photographic image—the Picture of the Century, as it is being called, a photographic montage of the planets from the perimeter of the solar system. The 60 pictures were stored in Voyager's recorder. In March and April, when ground stations were free to receive them, the images were transmitted back to Earth. The Jet Propulsion Lab expects to have the picture ready to release next month.

As a piece of hard data, the Picture of the Century is somewhat redundant; we've had a good fix on the position of most of the planets for 400 years. But its cultural value, its poetic importance, if you will, is enormous. From the edge of the solar system, the earth will seem to be a piece of cosmic dust, one of many small white specks suspended against a background of black, anonymous and sobering. And the fact that we have managed to get ourselves—or the 1,800-pound parcel of electronic and mechanical instruments that represents us—into a position to take that view gives a working scientist such as Ed Stone the opportunity to speak as a visionary.

"I keep asking myself why is there so much public interest in space," he says. "It is, after all, just basic science, in the end. The answer is that it provides us with a sense of the future. When we stop discovering new things out there, the concept of the future will change. Space reminds us that there is something left to be done, that life will continue to evolve. It gives us direction, an arrow in time."

As his work on Voyager comes to a close, Stone is not slackening his pace; he is either leading or taking part in four other space projects. If not the country's most prolific space scientist, he is probably its busiest.

In fact, his professional plate has always been full. For 26 years, he has taught and conducted research at the California Institute of Technology, a premier scientific center; he is currently vice president for astronomical facilities there. Since 1972, he has been chief scientist on the Voyager Project at the Jet Propulsion Lab, the arm of the National Aeronautics and Space Administration that has built and monitored many unmanned spacecraft. For the last two years, he has also acted as head operations officer of a $94 million partnership, between Caltech and the University of California, that is building the world's largest land-based telescope, the Keck Observatory, nearing completion on Mauna Kea, in Hawaii.

At this particular moment, Stone is sipping tea from a brown plastic cup, which one of his secretaries reckons has not had a proper washing in two decades. He is seated at a long table in a windowless conference room, presiding at the regular weekly meeting of his Caltech research team. The group is engaged in building a cosmic-ray spectrometer for a 390-pound, $23 million NASA science spacecraft called the Solar Anomalous and Magnetospheric Particle Explorer.

Sampex, a three-year mission that will be launched in 1992, is a package of three devices designed to measure and study subatomic particles from the sun and from space, clues to the origin of matter. The spacecraft is part of NASA's Small Explorers Program, an effort by the agency to launch three small research payloads between 1992 and 1994. These modest unmanned flights represent a tiny fraction of the space budget—this fiscal year roughly $13 million out of $12.3 billion—but they are vital to scientists who specialize in the physics of space and must rely on the Government to get their apparatus aloft and into the field.

There is a visitor at this meeting—Timothy C. Gehringer, an instrument coordinator for the Sampex mission, one of NASA's men, who has come from Washington to announce that the administration of the Small Explorers Program has been revamped. No more red tape, he promises. To expedite the program, NASA is giving the scientists more autonomy. "There won't be 10 hands at headquarters asking you for information." From now on, he tells the Caltech men, they will be free to build their "box" as they see fit. "If you want to use Radio Shack parts, that's up to you," he says. "Basically, it's your reputations that are on the line here."

"Well, we're not going to really use Radio Shack parts," Stone explained later. "There are two approaches to putting experiments on spacecraft. One is to be held accountable for every detail of your device, to be made to document everything. The other approach is called the black box; we build the device and simply turn it over to NASA to put on the orbiter. Gehringer was saying what's really important is that we're building this device to do science. We don't get our return until we get the science."

Ed Stone has been putting experiments on spacecraft for almost 30 years—as long, one guesses, as many NASA men have been in long pants. He is today one of his profession's most productive experimentalists in space; the three missions he is involved in, in addition to Sampex and Voyager's long-range interstellar assignment, are each designed to address specific scientific questions:

On the Galileo project, launched last year, Stone is leading a team employing a heavy-ion counter to study the radiation belts of Jupiter. On the Advanced Composition Explorer, scheduled for the middle of the decade, for which he is also chief scientist, the six high-resolution spectrometers will analyze three kinds of subatomic particles: matter cast out from the sun, the material of local interstellar space, and high-energy particles from elsewhere in the galaxy. Finally, Stone is one of a number of people working on a powerful superconducting magnet spectrometer for America's proposed $16.5 billion space station. This device, called Astromag, will allow scientists to study a wide range of particles that previous instruments could either not capture or measure with precision.

In a sense, a cosmic-ray physicist is a deep-space anthropologist. Cosmic rays are not rays at all, but subatomic particles generated by solar eruptions and the explosion of stars. The particles are scattered through the universe, and cosmic-ray physicists spend their lives shagging and studying them.

The particles—isotopes of hydrogen, helium, magnesium, silicon and the other elements—are like artifacts and fossils; their number, composition and possible origin could force theorists to revise current ideas about the Big Bang and the cosmic events believed to have followed it: the blast of supernovae, the birth of planets.

The instruments that Ed Stone helps to design and build are made, at the core, of thin silicon wafers and disks. When a cosmic-ray particle passes through the layers of silicon, it produces electrical signals, signatures of its nuclear charge and velocity; from these it is possible to calculate its mass, literally its identity.

Thus, through their space experiments, Ed Stone and his colleagues are asking some fundamental questions: What is the basic composition of the sun and how does it differ from other stars, those still burning and those that turned to dust billions of years ago? What is the galaxy made of? And what elements lie beyond?

"If we can begin to understand the differences, then we'll have some clues to the basic evolution of the universe," Stone said.

For the most part, experimentalists such as Ed Stone perform in relative obscurity. Their work is arcane and rarely attracts popular notice. Usually, they are part of a large enterprise, such as a space shot, where the significance of their labor is lost in the effort of the group.

Often on space shots, teams come into conflict: a radio team might want one telemetry to measure a planet's mass while an imaging team wants another for high-resolution pictures. For many years—through the flights of Explorer, Ranger, Surveyor, Mariner and Viking—the scientific rivalries at the Jet Propulsion Lab often led to turf wars, personal feuds and shouting matches. "It simply was chaos," said one lab official. On Voyager, too, teams formed around a variety of experiments—11 in all, covering the general fields of radio science, ultraviolet astronomy, plasma-wave physics, infrared observation, imaging, cosmic rays, low-energy particles and magnetic fields. But instead of putting a professional administrator in charge of the Voyager teams, lab officials, in 1972, reached into the ranks and elevated Ed Stone, a cosmic-ray physicist, to be the project's chief scientist.

"Ed was a *real* scientist and not just someone filling a job slot," said Torrence V. Johnson, a member of Voyager's imaging team and now a head scientist himself, on Project Galileo.

Stone had the right profile for the job. He was a protégé of John Simpson, the University of Chicago physicist who had been doing high-altitude and space fieldwork since the late 1950's, one of the grand old men of space, and had himself been putting experiments on spacecraft since the 1961 flight of Discoverer 36.

He was considered a scholar: "If a question came up in a field he didn't know about, Ed would go home and read all the papers in that field, then come back the next day better informed than the so-called experts," said Johnson. And a workhorse as well: "He was like this machine," recalls Norman Haynes. "You'd wind him up and, zoom! He went zipping around all day getting things done."

Most important, he found a way to end the feuding. He insisted on consensus. If a team wanted time on line during an encounter—and time was limited; Voyager spent just five days and five hours close to Neptune—a team leader not only had to persuade Stone his case was compelling, he had to win over his peers on the other teams as well.

"Ed would ask, 'Who's right scientifically?'" Johnson said.

In short, as Haynes has said, "Ed Stone revolutionized the world of project science."

The process of science—forming a hypothesis, conducting the experiment, analyzing and interpreting the data, then offering it for a long period of peer review—often makes conservatives out of those who practice it. At first, Ed Stone had trouble convincing his cautious colleagues on the Voyager project that big science, publicly financed science, sometimes demands immediate results.

"I explained that in my mind we were observing nature as it displayed itself, not doing controlled lab experiments," he said. "The public was paying for all this, we had to allow them to share what they are paying for."

He wanted his principal investigators to speak at the daily news conferences during each encounter, even though they often were puzzled about much of the data they were collecting. Some felt Stone was advocating "instant science," and, as Torrence Johnson remembers it, many of them "had to be dragged kicking and screaming to the microphone."

"I told them," said Stone, "you just have to be careful not to explain things you can't explain. Explain only what you think you really know. You say, 'Look, it's a mystery; we don't know what the data is telling us.' This was not the normal scientific process where things are worked over by peer review, but I felt with these encounters there was a real opportunity to share the whole scientific process in a compressed form and we should do it even if it meant violating some of the commonly accepted rules of how you do science."

In fact, the Voyager mission revealed just how little scientists really know about the solar system.

"The old saying is that after you've done the thing, you have more questions than when you started? Well, that's what happened with Voyager," Stone said. "For example, one of the key questions we can now ask is how do you generate a magnetic field inside a planet. There is no complete theory. Before Uranus and Neptune, our experience was limited. All the magnetic fields we had seen were near the rotational poles of the planets. Then we fly by Uranus and find out to our great surprise that the magnetic axis was tilted 60 degrees from the rotational axis. At Neptune, we found another titled magnetic field. We asked: what's going on inside those planets? That's one of the key questions we did not have before. We thought we understood about all the magnetic fields."

The success of the 13-year mission also underscores a continuing debate within the United States space program. "Generally, space scientists do not share the opinion that the man-in-space program is important," said Stone. "They feel a lot more can be done with robotic probes. Science is not the reason you put the people in space. The motive is exploration. In fact, the complexity of the system is magnified greatly when you have to support human life. There is so much science you can do at a much lower cost. The concern many space scientists have is that the cost of keeping man in space will be so large there will be great sacrifices in the robotic program, which has stayed at 20 percent of NASA's budget for years."

Stone thinks of Voyager as a laboratory bench, not a scout cutting a path to deep space. "The science is the excitement, not the getting there, not the exploration, but the chance to be at the lab bench. To a scientist, it's, 'Let's turn this thing on and see what we get. Let's experiment.'"

Edward C. Stone Jr., 54 years old, grew up in a small one-story house in Burlington, Iowa. He pursued physics from the time he was in high school, and since the fall of 1957, when the Russians launched Sputnik and opened a new age, he has wanted to experiment in space.

In those years, Stone was studying at the University of Chicago, working with physicists such as Simpson and Eugene N. Parker, a theorist. Simpson shaped his intellect, Parker his imagination. "I used to eat a brown-bag lunch with Parker in his lab once a week," Stone said. "We would speculate about what things were like in space—about the sun and cosmic rays, magnetic fields and their atmospheres. Parker taught me how to reduce a problem to its nuts and bolts, to a picture."

Meanwhile, on a blind date at a comedy club, he met Alice Wickliffe, an undergraduate from Camp Point, Ill. They married in 1962, arrived in Southern California two years later and, eventually, with two daughters, came to settle on a quiet street in Altadena in a green, one-story stucco bungalow with hibiscus and oleander and Italian cyprus in front.

Over the years, Ed Stone has been a shy man, sometimes diffident, often detached, say some who have worked at his side. "People realize his time is valuable, so they don't do idle chit-chat," says Alan C. Cummings, a scientist in Stone's research group at Caltech. "It's physics, and we don't bother with, 'How's the wife and kids?'"

This quality has, in some eyes, made Stone into a riddle. "Unlike some other people you get to know well, it was difficult to get to know Ed," says George Gloeckler, a classmate from Stone's days at Chicago and now a professor at the University of Maryland. "He never let his hair down; you never knew what he really thought."

To make too much of this, however, is misleading, for Stone is not in reality at all distant. And to illustrate this, Carolynn Young, his former assistant on Voyager, offers the following story: "We had this young person on the project and he was an obnoxious jerk. Ed said, 'I think you guys are being unkind and unfair. These kids are intellectually gifted, but often socially inept. A lot of their time is spent in books, and they don't know any better.' Well, the young person still works here and is still a jerk, but the incident made me wonder if Ed had been one of those kids. You see, there's a warmth there."

Stone's younger daughter, Janet, who is 25, says she has never seen him sulk or lose his temper. "I don't mean to make him sound like Mr. Spock," she says, "but he's probably the most in control of anyone I know." He listens to Mozart and Vivaldi and used to play the French ho rn. A registered Democrat, he's "not really political" (this is said as if something sour has just crossed his tongue). He doesn't have a best friend "in the classic sense." It would be more accurate to say he has shared a long professional kinship with several scientists. He goes home for dinner every night, but frequently returns to the institute to work. He has no interest in sports, no hobbies, really, though he has adopted his wife's enthusiasm for architecture. His main recreation is to read a daily newspaper. His favorite food is raisin pie. He is not a man of faith. "I'm comfortable with religion," he says, "but I don't have a need for a religion."

His values are the values of science. Whatever the job—a problem, a project, a large enterprise—in the end it must be "interesting" or he will not embrace it.

"Interesting" is a physicist's favorite superlative. It means: compelling, engaging, but most of all, worthy; in other words, the task must be deserving of the time and effort of a trained mind. As Eugene Parker, professor of physics at the University of Chicago, explains: " I might get curious about sunspots and do a calculation and say, 'Ah, that's an interesting idea.' To solve a problem, a scientist must have his mind occupied; he must be interested."

For a physicist, nothing is more "interesting" than nature. And because there is so much in nature that is unknown, the mind of an interested physicist, such as Ed Stone, is usually occupied. It therefore should come as little surprise when he declares, "My work's my hobby." In truth, it's his life.

And then there's Voyager's final mission—to reach and mark the heliopause. This frontier, if it may be called such, is marked by the meeting of two different winds. The sun sends out so many subatomic particles they form a "solar wind." The velocity of the solar wind is fierce, roughly one million miles per hour, but at a point beyond the orbits of the two most distant planets— Neptune or Pluto, depending on their changing paths—the solar wind slows in the face of another cosmic stream, the interstellar wind, and begins to ebb. At some farther point, the force of solar wind actually yields to the force of interstellar wind. This point is called the heliopause, and it is where the sun's influence truly ends. Here indeed is the frontier, the end line of the solar system.

Cosmic rays travel at two different speeds. High-energy particles are fast enough that they have the power to traverse the solar wind; they have been captured in our solar system. Low-energy particles lack this power, and so far have eluded study. The only way to study low-energy particles—and learn something of the ancient stars that produced them—is to get a cosmic-ray spectrometer, such as the one on Voyager, beyond the heliopause.

Scientists guess that the frontier may be some 120 astronomical units—or more than 11 billion miles—from the sun. This distance, and the time it takes to traverse it, gives the Voyager story a curious twist.

The mission may not be completed for at least 20 years more. "Now that you know that," says Torrence Johnson, "think about this: Ed Stone's experiment on Voyager, the cosmic-ray experiment, hasn't achieved its results yet—which was to get out beyond the heliopause and measure soft cosmic rays before the solar wind and the sun's gravity field could scatter and absorb them."

In other words, the chief scientist on Project Voyager has spent the better part of his career guiding a project that will contribute to his life's work only after he's retired or dead.

"I know there's a certain irony to my situation," says Ed Stone, "but I've been having so much fun on Voyager that even if I never see the edge of the solar system, I would do it all again."

"Someday," he goes on, "someday, somebody will make it to the interstellar medium.

"You take it one step at a time."

IV. OUT OF THE CRADLE

EDITOR'S INTRODUCTION

Why should the United States have a national space program, and what should such a program accomplish? We have tried to accomplish many goals in space, some of which have conflicted with one another. For example, the consensus among space scientists is that NASA should concentrate on the less expensive, less complicated robotic probes as opposed to the man-in-space missions. The manned space missions satisfies the public's sense of adventure in exploring the unknown but maintaining human life in space introduces a whole new dimension of complexity. The fundamental problem is that there has been no clear agenda for the space program since President Kennedy's 1961 speech in which he promised that we would have a man on the moon before the end of the decade. Now, twenty one years after Neil Armstrong set foot on the moon's surface and uttered his famous words, the United States has no way of returning to the moon; the giant Saturn boosters of the Apollo missions have been dismantled or stand rusting in "rocket gardens." Even their blueprints have been lost or destroyed. Today's space program needs a more comprehensive and long-term vision. More than new launch systems or space probes, the space agency needs to articulate clearly the necessity for spaceflight, the need for human exploration of space, and the kinds of missions that will yield the best science, the greatest potential for commercialization, and the highest adventure. In the first article former astronaut Sally K. Ride offers her assessment of why we should be in space, and where we should go from here, in her influential report to the National Research Council, "Leadership and America's Future in Space."

Will space travel ever become commonplace? Radical new technologies will be required before someone can hop a flight into orbit. Such technology is already being developed. In the second selection, George Keyworth, former science advisor to President Reagan, and Bruce Abbell, describe a space plane that will take off and land on conventional runways at a launch cost millions of dollars less than that of the space shuttle. Such a space plane could make space tourism a reality, claims John Dennison in "Getting a Lift to LEO" for *Ad Astra*. Potential passengers have already made down payment on the maiden voyage to low earth orbit, even though no one is certain when such a feat will be possible. Beyond earth orbit beckon the Moon and Mars. The next article discusses international plans to establish a permanent lunar base and to visit the red planet which is more earthlike than all the other planets in our solar system. The multibillion dollar price tags for these missions—as much as $400 billion for an expedition to Mars—make international cooperation for future space efforts essential. The end of the Cold War and the increasing globalization of technology makes this a better prospect than ever before.

Despite the allure of space travel, not everyone is convinced that it is worth the effort. We are busily making a shambles of our own world, critics claim, and we're likely to do the same to the rest of the universe. In the final piece, James Marti sums up several dissenting views from small journals. As one writer notes, "Without a thorough overhaul of human values before we get there, the colonization of space will simply transport our 'planetary hangups' somewhere else."

LEADERSHIP AND AMERICA'S FUTURE IN SPACE[1]

For nearly a quarter of a century, the U.S. space program enjoyed what can appropriately be termed a "golden age." From the launch of Earth-orbiting satellites, to the visits by robotic spacecraft to Venus and Mars, to the stunning achievement of landing the first human beings on the Moon, the many successes of the space program were exciting and awe-inspiring. The United States was clearly and unquestionably the leader in space exploration, and the nation reaped all the benefits of pride, international prestige, scientific advancement, and technological progress that such leadership provides.

[1]Abridged from a report to NASA by former astronaut Sally K. Ride, published by the National Research Council in 1987.

However, in the aftermath of the *Challenger* accident, reviews of our space program made its shortcomings starkly apparent. The United States' role as the leader of the spacefaring nations came into serious question.

The U.S. civilian space program is now at a crossroads, aspiring toward the visions of the National Commission on Space but faced with the realities set forth by the Rogers Commission. NASA must respond aggressively to the challenges of both while recognizing the necessity of maintaining a balanced space program within reasonable fiscal limits. . . .

The goals of the civilian space program must be carefully chosen to be consistent with the national interest and also to be consistent with NASA's capabilities. NASA alone cannot set these goals, but NASA must lead the discussion, present technically feasible options, and implement programs to pursue those goals which are selected.

We must ask ourselves: "Where do we want to be at the turn of the century?" and "What do we have to do now to get there?" Without an eye toward the future, we flounder in the present. . . .

Leadership in space does not require that U.S. be preeminent in all areas of space enterprise. The widening range of space activities and the increasing number of spacefaring nations make it virtually impossible for any country to dominate in this way. It is, therefore, essential for America to move promptly to determine its priorities and to pursue a strategy which would restore and sustain its leadership in the area deemed important. . . .

Being an effective leader does mandate, however, that this country have capabilities which enable it to act independently and impressively when and where it chooses, and that its goals be capable of inspiring others—at home and abroad—to support them. It is essential for this country to move promptly to determine its priorities and to make conscious choices to pursue a set of objectives which will restore its leadership status. . . .

In response to growing concern over the posture and long-term direction of the U. S. civilian space program, NASA Administrator Dr. James Fletcher formed a task group to define potential U.S. space initiatives, and to evaluate them in light of the current space program and the nation's desire to regain space and retain leadership. The objectives of the study were to energize a discussion of the long-range goals of the civilian space program and to begin to investigate overall strategies to direct that program to a position of leadership.

The task group identified four candidate initiatives for study and evaluation. Each builds on NASA's achievements in science and exploration, and each is a bold, aggressive proposal which would, if adopted, restore the United States to a position of leadership in a particular sphere of space activity. The four initiatives are: (1) Mission to Planet Earth, (2) Exploration of the Solar System, (3) Outpost on the Moon, and (4) Humans to Mars. All four initiatives were developed in detail, and the implications and requirements of each were assessed.

This process was not intended to culminate in the selection of one initiative and the elimination of the other three, but rather to provide four concrete examples which would catalyze and focus the discussion of the goals and objectives of the civilian space program and the efforts required to pursue them. . . .

A U.S. space leadership program must have two distinct attributes. First, it must contain a sound program of scientific research and technology development—a program that builds the nation's understanding of space and the space environment, and that builds its capabilities to explore and operate in that environment. The United States will not be a leader in the 21st Century if it is dependent on other countries for access to space or for the technologies required to explore the space frontier. Second, the program must incorporate visible and significant accomplishments; the United States will not be perceived as a leader unless it accomplishes feats which demonstrate prowess, inspire national pride, and engender international respect and a worldwide desire to associate with U.S. space activities. . . .

The next step in this process should be to articulate specific objectives and to identify the programs required to achieve these objectives. . . . Does this country intend to establish a lunar outpost? To send an expedition to Mars? What are NASA's major objectives for the late 20th and early 21st Centuries? The Space Shuttle and Space Station will clearly support the objectives, but what will they be supporting?

These questions cannot, of course, be answered by NASA alone. But NASA should lead the discussion, propose technically feasible options, and make thoughtful recommendations. The choice of objectives will shape, among other things, NASA's technology program, the evolution of the Space Station, and the character of Earth-to-orbit transportation. . . .

The ground rules for this study are important to understand, since they influenced the detailed definition of the initiatives. The ground rules, set forward at the outset of this study, were:

• The initiatives should be considered in *addition* to currently planned NASA programs. They were not judged against, nor would they supplant, existing programs.

• Each initiative should be developed independently. There is, of course, considerable synergism between certain initiatives. For example, one possible progression for human exploration could be the development of a lunar outpost, followed by an expedition to Mars. However, in order to provide a clear starting point for discussion, the four were considered to be distinct.

• The initiatives should achieve major milestones within two decades.

• The Humans to Mars initiative should be assumed to be an American venture. It was beyond the scope of this work to consider joint U.S./Soviet human exploration.

The candidate initiatives were developed and presented NASA management to: (1) evaluate the initiatives and their implications, and (2) promote a discussion of the attributes of each initiative to determine the elements which are most important to NASA and to the United States.

Mission to Planet Earth

Mission to Planet Earth is an initiative to understand our home planet, how forces shape and affect its environment, how that environment is changing, and how those changes will affect us. The goal of this initiative is to obtain a comprehensive scientific understanding of the entire Earth System, by describing how its various components function, how they interact, and how they may be expected to evolve on all time scales. . . .

With the launch of the first experimental satellites in the 1960s, NASA pioneered the remote sensing of Earth from space. Over the past two decades, the scientific community has concluded that Earth is in a process of global change, and scientists now believe that it is necessary to study Earth as a synergistic system. . . . Interactive physical, chemical, and biological processes connect the oceans, continents, atmosphere, and biosphere of Earth in a complex way. Oceans, ice-covered regions, and the atmosphere are closely linked and shape Earth's climate; volcanism links inner Earth with the atmosphere; and biological activity significantly contributes to the cycling of chemicals (e.g., carbon, oxygen, and carbon dioxide) important to life. And now it is clear that human activity also has a major impact on the evolution of the Earth System.

Global-scale changes of uncertain impact, ranging from an increase in the atmospheric warming gases, carbon dioxide and methane, to a hole in the ozone layer over the Antarctic, to important variations in vegetation covers and in coastlines, have already been observed with existing measurement capabilities. The potentially major consequences, either detrimental or beneficial, suggest an urgent need to understand these variations.

We currently lack the ability to foresee changes in the Earth System, and their subsequent effects on the planet's physical, economic, and social climate. But that could change; this initiative would revolutionize our ability to characterize our home planet, and would be the first step toward developing predictive models of the global environment.

The guiding principle behind this initiative is to adopt an integrated approach to observing Earth. The observations from various sensors on platforms and satellites will be coordinated to perform global surveys and also to perform detailed observations of specific phenomena.

Mission to Planet Earth proposes:

1. To establish and maintain a global observational system in space, which would include experiments and free-flying platforms, in polar, low-inclination, and geostationary orbits, and which would perform integrated, long-term measurements.

2. To use the data from these satellites along with *in-situ* information and numerical modeling to document, understand, and eventually predict global change.

The global observational system would include a suite of nine orbiting platforms:

• Four sun-synchronous polar platforms: two provided by the United States and one each provided by the European Space Agency (ESA) and the Japanese National Space Development Agency (NASDA). The first platform would be launched in 1994 and all four platforms would be in orbit by 1997. These platforms would provide global polar coverage with morning and afternoon crossing times.

• Five geostationary platforms: three provided by the U.S. and one each by ESA and NASDA. These platforms would all be launched and deployed between 1996 and 2000. . . .

The integrated system would measure the full complement of the planet's characteristics, including: global cloud cover, vegetation cover, and ice cover; global rainfall and moisture; ocean chlorophyll content and ocean topography; motions and deformations of Earth's tectonic plates; and atmospheric concentration of gases such as carbon dioxide, methane, and ozone.

Space-based observations would also be coordinated with ground-based experiments and the data from all observations would be integrated by an essential component of this initiative: a versatile, state-of-the-art information management system. This tool is critical to data analysis and numerical modeling, and would enable the integration of all observational data and the development of diagnostic and predictive Earth System models. . . .

To achieve its full scope, this initiative requires the operational support of Earth-to-orbit and space transportation systems to accommodate the launching of polar and geostationary platforms. This does not represent a large number of additional launches, but it does require the capability to launch large payloads to polar orbit; Titan IVs would be used to accomplish this. Since the envisioned geostationary platforms would be lifted to low Earth orbit, assembled at the Space Station, and then lifted to geosynchronous orbit with a space transfer vehicle well-developed orbital facilities are essential. By the late 1990s, the Space Station must be able to support on-orbit assembly, and a space transfer vehicle must exist. . . .

Mission to Planet Earth is not the sort of major program the public normally associates with an agency famous for *Apollo, Viking,* and *Voyager.* But this initiative is a great one, not because it offers tremendous excitement and adventure, but because of its fundamental importance to humanity's future on this planet. . . .

The U.S. is the only country currently capable of leading a Mission to Planet Earth, but the program is designed around, and requires, international cooperation. Admittedly, the initiative's international scope could complicate its coordination and implementation, but the concept embodied in the initiative enjoys widespread international support. As more and more countries are facing ecological problems, there is increasing interest in a global approach. . . .

NASA should embrace Mission to Planet Earth. This initiative is responsive, time-critical, and shows a recognition of our responsibility to our home planet. Do we dare apply our capabilities to explore the mysteries of other worlds, and not also to apply those capabilities to explore and understand the mysteries of our own world—mysteries which may have important implications for our future on this planet?

Exploration of the Solar System

This initiative would build on NASA's long-standing tradition of solar system exploration and would continue the quest to understand our planetary system, its orgin, and its evolution. Solar system bodies are divided into three distinct classes: the primitive bodies (comets and asteroids), the outer (gas giant) planets, and the inner (terrestrial) planets. Each class occupies a unique position in the history of the solar system, and each is a target of a major mission in its initiative, which includes a comet rendezvous (the Comet *Rendezvous Asteroid Flyby* mission), a mission to Saturn (*Cassini*), and three sample return missions to Mars. The centerpiece of the initiative is the robotic exploration of Mars; the first of these three automated missions would bring a handful of Mars back to Earth before the year 2000. . . .

Although currently schedueld U.S. missions will ensure that the United States will remain a leader in certain areas of solar system exploration through 1995, the position of the United States beyond 1995 is in question. This initiative would maintain U.S. leadership in exploration of the outer planets, and would regain and sustain U.S. leadership in the exploration of both the planet Mars and the primitive bodies of the solar system.

This initiative is based on the balanced strategy developed by the Solar System Exploration Committee of the NASA Advisory Council. . . . The missions include:

1. The *Comet Rendezvous Asteroid Flyby (CRAF)* mission would investigate the beginnings of our solar system, studying a Main Belt asteroid and a comet, which represent the best-preserved samples of the early solar system. Because of their primordial nature, comets can provide critical clues about the processes that led to the origin and evolution of our solar system.

The *CRAF* mission scenario [begins with] a 1993 launch and a six-month cruise. The spacecraft would fly past the asteroid Hestia at an altitude of about 10,000 kilometers. *CRAF*'s visual and infrared asteroid imaging systems would conduct investigations of Hestia's surface composition and structure. *CRAF* would then continue its journey for a rendezvous with a periodic comet, Tempel 2. The spacecraft would maneuver to within 25 kilometers of the comet's nucleus and begin a series of observations, which includes shooting two penetrators into the nucleus itself for detailed *in-situ* measurements. The spacecraft would fly in close formation with the comet until it nears the Sun and becomes active; then the spacecraft would maneuver farther away to observe the comet's coma and tail.

2. The *Cassini* mission would explore Saturn and its largest moon, Titan. The giant outer planets offer us an opportunity to address key questions about their internal structures and compositions through detailed studies of their atmospheres. Titan is an especially interesting target for exploration because the organic chemistry now taking place there provides the only planetary-scale laboratory for studying processes that may have been important in the prebiotic terrestrial atmosphere.

The *Cassini* mission proposed in this initiative would be a considerably expanded version of the *Cassini* mission considered by the Solar System Exploration Committee. This expanded mission would be launched in 1998 for the long interplanetary voyage to arrive at Saturn in 2005 with a full array of investigative instruments. An orbital spacecraft and three probes would conduct a comprehensive three-year study of the planet and its rings, satellites, and magnetosphere. One atmospheric probe would be launched toward Titan. The expanded *Cassini* mission would also carry one probe to investigate the Saturnian atmosphere, and one semi-soft lander which would reach the surface of Titan.

3. The *Mars Rover/Sample Return* mission would, in journeys covering hundreds of millions of miles, gather samples of Mars and bring them back to Earth. Because of its relevance to understanding Earth and other terrestrial planets, and because it is the only other potentially habitable planet in our solar system, Mars is an intriguing target for exploration.

The *Mars Rover/Sample Return* mission scenario would involve a soft landing on the Martian surface deployment of a "smart" surface rover to select and collect samples, delivery of the samples to an ascent vehicle, and transfer of the samples from Mars orbit to a return vehicle. The samples would then most likely be returned to a sample-handling module on the Space Station for analysis.

The initiative would include three such missions: two launched in 1996, probably sending redundant rovers and ascent vehicles to ensure return of a sample in 1999, and one launched in 1998/99 with return in 2001.

As it is defined, this initiative places a premium on advanced technology and enhanced launch capability to maximize the scientific return. It requires aerobraking technology for aerocapture and aeromaneuvering at Mars, and a high level of sophistication in automation, robotics, and sampling techniques. Advanced sampling methods are necessary to ensure that geologically and chemically varied and interesting samples are collected for analysis.

The Solar System Exploration initiative significantly benefits from improved launch capability in terms of the science returned from both the Mars and *Cassini* missions. In fact, it is a heavy-lift launch vehicle that enables the full complement of three different probes to be carried in the expanded *Cassini* mission.

The Space Shuttle is not required for any of the missions in the initiative. The Space Station would not be needed until 1999, when an isolation module may be used to receive the Martian samples. . . .

Although the *Mars Rover/Sample Return* was presented as a U.S. mission in this initiative, it could be performed in cooperation with our allies and/or in coordination with the Soviet Union. . . .

Robotic planetary exploration should be actively supported and nurtured within NASA. Although it does not have the immediate relevance of the Mission to Planet Earth, or the excitement of human exploration, it is fundamental science that challenges our technology, extends our presence, and gives a glimpse of other worlds. As such, it enjoys widespread public interest and support. Although not necessarily at the pace suggested in this initiative, planetary exploration must be solidly supported through the 1990s.

Outpost on the Moon

This initiative builds on the legacy of *Apollo* and envisions a new phase of lunar exploration and development—a phase leading to a human outpost on another world. That outpost would support scientific research and exploration of the Moon's resource potential, and would represent a significant extraterrestrial step toward learning to live and work in the hostile environments of other worlds.

Beginning with robotic exploration in the 1990s, this initiative would land astronauts on the lunar surface in the year 2000, to construct an outpost that would evolve in size and capability and would be a vital, visible extension of our capabilities and our vision. . . .

This initiative represents a sustained commitment to learn to live and work in space. As our experience and capabilities on the lunar surface grow, this extraterrestrial outpost will gradually become less and less dependent on the supply line to Earth. The first steps toward "living off the lunar land" will be learning to extract oxygen from the lunar soil, where it is plentiful, and learning to make construction materials. The lunar soil would eventually be a source of oxygen for propellant and life-support systems, and a source of material for shelters and facilities.

The Moon's unique environment provides the opportunity for significant scientific advances; the prospect for gains in lunar and planetary science is abundantly clear. Additionally, since the Moon is seismically stable and has no atmosphere, and since its far side is shielded from the radio noise from Earth, it is a very attractive spot for experiments and observations in astrophysics, gravity wave physics, and neutrino physics, to name a few. It is also an excellent location for materials science and life science research because of its low gravitational field (one-sixth of Earth's).

This initiative proposes the gradual, three-phase evolution of our ability to live and work on the lunar surface.

Phase I: Search for a Site (1990s). The initial phase would focus on robotic exploration of the Moon. It would begin with the launching of the *Lunar Geoscience Observer*, which will map the surface, perform geochemical studies, and search for water at the poles. Depending on the discoveries of the *Observer*, robotic landers and rovers may be sent to the surface to obtain more information. Mapping and remote sensing would characterize the lunar surface and identify appropriate sites for the outpost. . . .

Phase II: Return to the Moon (2000–2005). Phase II begins with the return of astronauts to the lunar surface. The initiative proposes that a crew be transported from the Space Station to lunar orbit in a module propelled by a lunar transfer vehicle. The crew and equipment would land in vehicles derived from the transfer vehicle. Crew members would stay on the surface for one or two weeks, setting up scientific instruments, a lunar oxygen pilot plant, and the modules and equipment necessary to begin building a habitable outpost. The crew would return to the orbiting transfer vehicle for transportation back to the Space Station.

Over the first few flights, the early outpost would grow to include a habitation area, a research facility, a rover, some small machinery to move lunar soil, and a pilot plant to demonstrate the extraction of lunar oxygen. By 2001, a crew could stay the entire lunar night (14 Earth days), and by 2005 the outpost would support five people for several weeks at a time.

Phase III: At Home on the Moon (2005–2010). Phase III evolves directly from Phase II, as scientific and technological capabilities allow the outpost to expand to a permanently occupied base. The base would have closed-loop life-support systems and an operational lunar oxygen plant, and would be involved in frontline scientific research and technology development. The program also requires the mobilization of disciplines not previously required in the space program: surface construction and transportation, mining, and materials processing.

By 2010, up to 30 people would be productively living and working on the lunar surface for months at a time. Lunar oxygen will be available for use at the outpost and possibly for propellant for further exploration.

This initiative envisions frequent trips to the Moon after the year 2000—trips that would require a significant investment in technology and in transportation and orbital facilities in the early 1990s. . . .

The transportation system must be capable of regularly transporting the elements of the lunar outpost, the fuel for the voyage, and the lunar crew to low-Earth orbit. This requires a heavy-lift launch vehicle and a healthy Space Shuttle fleet. The transfer of both cargo and crew from the Space Station to lunar orbit requires the development of a reusable space transfer vehicle. This and a heavy-lift vehicle will be the workhorses of the Lunar initiative.

The Space Station is an essential part of this initiative. As the lunar outpost evolves, the Space Station would become its operational hub in low-Earth orbit. Supplies, equipment, and propellants would be marshalled at the Station for transit to the Moon. It is therefore required that the Space Station evolve to include spaceport facilities.

In the 1990s, the Phase 1 Space Station would be used as a technology and systems test bed for developing closed-loop life-support systems, automation and robotics, and the expert systems required for the lunar outpost. The outpost would, in fact, rely on the Space Station for many of its systems and subsystems, including lunar habitation modules which would be derivatives of the Space Station habitation/laboratory modules. . . .

The establishment of a lunar outpost would be a significant step outward from Earth—a step that combines adventure, science, technology, and perhaps the seeds of enterprise. Exploring and prospecting the Moon, learning to use lunar resources and work within lunar constraints, would provide the experience and expertise necessary for further human exploration of the solar system.

The Lunar initiative is a major undertaking. Like the Mars initiative, it requires a national commitment that spans decades. It, too, demands an early investment in advanced technology, Earth-to-orbit transportation, and a plan for Space Station evolution. . . .

However, this initiative is quite flexible. Its pace can be controlled, and more important, adapted to capability. It is possible to lay the foundation of the outpost in the year 2000, then build it gradually, to ease the burden on transportation and Space Station at the turn of the century.

The Lunar initiative is designed to be evolutionary not revolutionary. Relying on the Space Station for systems and subsystems, for operations experience, and for technology development and testing, it builds on and gradually extends existing capabilities. Many of the systems needed for reaching outward to Mars could be developed and proven in the course of work in the Earth–Moon region. It is not absolutely necessary to establish this stepping stone, but it certainly makes sense to gain experience, expertise, and confidence nearer Earth first, and then to set out for Mars.

Humans to Mars

This bold initiative is committed to the human exploration, and eventual habitation, of Mars. Robotic exploration of the planet would be the first phase and would include the return of samples of Martian rocks and soil. Early in the 21st Century, Americans would land on the surface of Mars; within a decade of these first piloted landings, this initiative would advance human presence to an outpost on Mars. . . .

This leadership initiative declares America's intention to continue exploring Mars, and to do so not only with spacecraft and rovers, but also with humans. It would clearly rekindle the national pride and prestige enjoyed by the U.S. during the *Apollo* era. Humans to Mars would be a great national adventure; as such, it would require a concentrated massive national commitment—a commitment to a goal and its supporting science, technology, and infrastructure for many decades.

This initiative would:

1. Carry out comprehensive robotic exploration of Mars in the 1990s. The robotic missions would begin with the *Mars Observer*, include an additional *Observer* mission, and culminate in a pair of *Mars Rover/Sample Return* missions. These missions would perform geochemical characterization of the planet, and complete global mapping and support landing site selection and certification.

2. Establish an aggressive Space Station life sciences research program to validate the feasibility of long-duration spaceflight. This program would develop an understanding of the physiological effects of long-duration flights, of measures to counteract those effects, and of medical techniques and equipment for use on such flights. An important result would be the determination of whether eventual Mars transport vehicles must provide artificial gravity.

3. Design, prepare for, and perform three fast piloted round-trip missions to Mars. These flights would enable the commitment, by 2010, to an outpost on Mars.

The Mars missions described in this initiative are one-year, round-trip "sprints," with astronauts exploring the Martian surface for two weeks before returning to Earth. The chosen scenario significantly reduces the amount of mass which must be launched into low-Earth orbit, and by doing so brings a one-year round trip into the realm of feasibility. This is accomplished by splitting the mission into two separate parts—a cargo vehicle and a personnel transport—and judiciously choosing the launch date for each.

The Mars cargo vehicle minimizes its propellant requirements by taking a slow, low-energy trip to Mars. The vehicle would be assembled in low-Earth orbit and launched for Mars well ahead of the personnel transport, and would carry everything to be delivered to the surface of Mars plus the fuel required for the crew's trip back to Earth.

The personnel transport would be assembled and fueled in low-Earth orbit, and would leave for Mars only after the cargo vehicle had arrived in Mars orbit. It would carry a crew of six astronauts, crew support equipment, and propellant for the outbound portion of the trip. Once in Mars orbit, it would rendezvous with the cargo vehicle, refuel, and prepare for descent to the surface. The landing party would spend 10 to 20 days on the Martian surface, and then re-rendezvous with the personnel transport for the trip back to Earth orbit. . . .

A significant, long-term commitment to developing several critical technologies and to establishing the substantial transportation capabilities and orbital facilities is essential to the success of the Mars initiative. . . . Even with separate cargo and personnel vehicles, and technological advances such as aerobraking, each of these sprint missions requires that approximately 2.5 million pounds be lifted to low-Earth orbit. (In comparison, the Phase 1 Space Station is projected to weigh approximately 0.5 million pounds). It is clear that a robust, efficient transportation system, including a heavy-lift launch vehicle, is required. . . .

The Phase 1 Space Station is a crucial part of this initiative. In the 1990s, it must support the critical life sciences research and medical technique development. It will also be the technology test bed for life-support systems, automation and robotics, and expert systems.

Furthermore, we must develop facilities in low-Earth orbit to store large quantities of propellant, and to assemble large vehicles. The Space Station would have to evolve in a way that would meet these needs. . . .

A successful Mars initiative would recapture the high ground of world space leadership and would provide an exciting focus for creativity, motivation, and pride of the American people. . . .

Any expedition to Mars is a huge undertaking, which requires a commitment of resources which must be sustained over decades. Their task group has examined only one possible scenario for a Mars initiative—a scenario designed to land humans on Mars by 2005. This time-scale requires an early and significant investment in technology; it also demands a heavy-lift launch vehicle, a larger Shuttle fleet, and a transportation depot at the Space Station near the turn of the century. This would require an immediate commitment of resources and an approximate tripling of NASA's budget during the mid–1990s. . . .

One alternative is to retain the scenario developed here, but to proceed at a more deliberate (but still aggressive) pace, and allow the first human landing to occur in 2010. This spreads the investment over a longer period, and though it also delays the significant milestones and extends the length of commitment, it greatly reduces the urgency for Space Station evolution and growth, and consequently for transportation capabilities as well.

We must pursue a more deliberate program; this implies that we should avoid a "race to Mars." There is the very real danger that if the U.S. announces a human Mars initiative at this time, it could escalate into another space race. Whether such a race is real or perceived, there would be constant pressure to set a timetable, to accelerate it if possible, and to avoid falling behind. Schedule pressures, as the Rogers Commission [Presidential commission which investigated the Challenger explosion] noted, can have a very real, adverse effect. . . .

Settling Mars should be our eventual goal, but it should not be our next goal. . . . We should adopt a strategy of natural progression which leads, step by step, in an orderly, unhurried way, inexorably toward Mars.

HOW TO MAKE SPACE LAUNCH ROUTINE[2]

Since the last moon landing in 1973, the momentum of the space shuttle has dominated U.S. space launch programs. Here was a vehicle, we were told, that would make access to space routine. It would launch satellites, carry supplies to build a space station, and provide a zero-gravity platform for scientific research and, ultimately, industrial production. During the early 1980s, our national policy relied almost solely on the shuttle as our means of reaching earth orbit.

This policy was doomed from the start. The shuttle is too costly, too complex, and too inflexible to support today's space access needs. Moreover, overreliance on any single system leaves us extremely vulnerable in the event of an accident; after the *Challenger* tragedy, U.S. access to earth orbit virtually disappeared for almost three years.

After 20 years, it is no surprise that a system begins to look dated and inadequate. But addressing these realities head-on has often been awkward, even painful, because so much money and effort has been invested in the current launch systems and because there is no replacement system immediately on the horizon.

But one research program now under way offers hope for precisely the kind of workaday access to space that shuttle proponents once envisioned: the National Aerospace Plane, or NASP. Unlike other launch vehicles that exist or are being developed, this aircraft would take off from a runway. It would then hurtle through the atmosphere at more than 20 times the speed of sound (Mach 20), deposit its payloads in low earth orbit, and finally descend and land on a runway.

NASP has been publicly perceived as primarily a hypersonic aircraft, intended for high-speed transport on earth, or as an exotic military reconnaissance or rapid deployment aircraft. (Conceived in the late 1970s by the Defense Advanced Research Projects Agency, NASP has been funded since 1985 through a joint NASA-Air Force program office at Wright-Patterson Air

[2]Article by George A. Keyworth, former director of the White House Office of Science and Technology Policy, and Bruce R. Abell, a senior research fellow at the Hudson Institute. Reprinted with permission from *Technology Review*, Copyright 1990.

Force Base.) When the program first entered the public spotlight, much was made of the idea of an "Orient Express" that could carry passengers from New York to Tokyo in one or two hours.

But the technology's most immediate impact will be in space access. More than any other launch vehicle now being considered, NASP would provide low-cost and flexible access to space. Unfortunately, plans for post-shuttle space-launch systems have not yet seriously included vehicles using NASP technologies.

The United States faces a serious shortage of lift capacity. Right now the U.S. fleet is able to launch about a million pounds a year into low earth orbit. That does not even adequately cover the "official" launch demand compiled by the Air Force Space Command.

And demand will surely increase in coming years. The need for communications satellites will continue to grow as direct-broadcast television services are put into place around the world. Further demand will come from a proliferation of satellite-based navigation and position-locating systems; a new generation of earth-sensing programs, including the ambitious Mission to Planet Earth; and a backlog of planetary exploration projects. Moreover, military need for space access will increase, not decrease, as international tensions ease and surveillance supplants readiness as the basis for national security. When nations reduce defenses, they put a higher premium on intelligence—the old adage trust but verify.

Many of these missions could be served by unmanned launch vehicles. But if the United States seriously wants to build a space station or explore Mars—both proposed national goals—it will need a way to get people as well as payloads into orbit cheaply. Interplanetary manned exploration in particular will be unrealistic unless we reduce the cost of access to space.

But more importantly, there is a huge class of potential users of earth orbit who cannot afford present launch systems. Each shuttle launch costs about $275 million, or $5,000 per pound of payload. Unmanned rockets are less expensive, but not dramatically; it costs about $150 million to put up a workhorse like the Titan, or about $3,000 per pound of payload. These costs form a high barrier to participation. Many more users will surface if prices drop to the $20 to $200 per pound range promised by NASP.

Providers of communications services other than for mass broadcast, for example, could take advantage of networks of satellites in low earth orbit (rather than in higher geosynchronous orbits). The idea of growing new kinds of materials in space—which so far has been essentially a stunt given the high cost of shuttle launches—could become economical.

Shuttle launches are not only expensive, but also infrequent. Routine access means frequent launches and it means the ability to launch on relatively short notice. Yet after 40 years of space access with rockets, the United States is still a long way from that capability, even for unmanned systems.

Another element of routine access is the ability to carry a wide range of payload sizes. The space shuttle is designed to carry one or at most two large payloads. But this practice is anachronistic. Very few things that we want to put into space are the size of the Hubble space telescope. Most projected flight requirements would require a payload of no more than 25,000 pounds.

This shift toward smaller payloads is becoming particularly apparent in the growing popularity of small, cheap satellites. Conventional satellites, with their billion-dollar price tags, are the space-borne equivalent of mainframe computers. Each multi-ton unit takes a decade (or longer) to build. Moreover, launch times must be reserved years in advance. Small satellites, typically weighing 50 to 1,000 pounds, are more like personal computers (PCs); they can be assembled quickly from inexpensive and accessible hardware and launched on short notice. And like the PC, as small satellites become more available, people will find unexpected new uses for them, further stimulating demand for launch services.

The move toward PC-equivalents in space is particularly important for scientists studying the earth. Missing from attempts to create realistic models of natural systems has been the ability to make large numbers of observations. We could learn more about the dynamics of global warming, for example, with 500 100-pound satellites than with one 50,000-pound satellite.

The shuttle's high cost and inflexibility have brought forth a number of proposals for alternative launch vehicles. But most of these would suffer the same deficiencies that mark the shuttle.

The most ambitious goals for reducing cost come from NASA's planned Advanced Launch System (ALS), an unmanned rocket that could carry cargoes of up to 200,000 pounds—more than the shuttle or its planned follow-on, the Advanced Manned Launch System. About half the cost of a shuttle launch goes not to hardware or to fuel but to operations; 12,000 people work on each launch. Developers of ALS aim to improve launch efficiency by simplifying operations and trimming this enormous personnel cost. Their target is a payload cost of about $300 per pound.

This order-of-magnitude reduction from the shuttle's cost is a laudable but not very credible goal. The ALS is, at its core, an extrapolation of well-mined technology. It shares the inherent inefficiencies of all multistage rockets—the costs of hauling expendable superstructure and liquid oxygen into space, for example. A recent study by the Office of Technology Assessment (OTA) projects an R&D cost on the order of $7 billion to develop the ALS, with an additional $4 billion for the new launch facilities that would be required to permit the more efficient operations. To pay back this large up-front costs, ALS would have to dominate the launch market—an unlikely prospect, given that ALS is intended to launch large satellites rather than the increasingly popular small ones.

In addition to NASA's projects, a number of commercial firms are promoting a class of small launch vehicles. One innovative example is the Pegasus, the air-launched version of which was successfully tested early this year. Like the old X-15 rocket plane, Pegasus is carried aloft under the wing of a large airplane and then drop-launched. The rocket expends no fuel until it reaches an altitude of perhaps 50,000 feet.

Small launchers such as the Pegasus or Space Services' Conestoga rocket are suitable for these smaller payloads, and satellite manufacturers are now designing and building families of small satellites sized to fit into these launch vehicles. Four of Ball Aerospace's Techstars, for example, can nest into the Pegasus housing. But lacking economies of scale, per-pound launch costs are still high—$10,000 to $15,000 for Pegasus, several times those of the shuttle.

The limitations of all these multistage, expendable launch vehicles have led some people to reexamine the idea of a launch system that climbs from the ground into orbit with a single rocket stage. The inherent difficulty of single stage to orbit, or SSTO, is the weight penalty for carrying the whole launch vehicle into orbit. For this reason, early rocket developers produced vehicles with expendable stages that are jettisoned after they use up their propellant. Over the years, engineers got comfortable with staged rockets, sometimes forgetting that a design decision based on technology of the 1950s and 1960s might be worth revisiting.

Technological developments are making nonexpendable SSTO launchers a more sensible alternative. Fuel tanks made from composite materials, for example, are much lighter than before. New designs also make possible lighter weight and more efficient rocket motors. Novel engine-control systems would continuously vary the fuel-to-oxidizer ratio to maintain the most efficient mixture as the vehicle climbs through the thinning atmosphere. This concept has recently been advanced by the work of a small number of innovators, notably Gary Hudson through his start-up, Pacific American Launch Systems (Redwood City, Calif.).

Developers of SSTO vehicles think they can substantially reduce launch costs. The Spaceship Experimental (SSX), for example, a vertical-takeoff and vertical-landing manned vehicle being developed by the Strategic Defense Initiative Office, aims for payload costs a fraction of the shuttle's. However, this manned, reusable vehicle is still a paper design.

Unlike the shuttle or any of the other proposed alternatives, a NASP vehicle would offer frequent and low-cost access to low earth orbit without the complexities of a rocket launch.

NASP vehicles will use a jet engine to fly to hypersonic speeds and high altitudes. Equipped with massive air intakes, they should ideally reach Mach 22 (about four miles per second) in air-breathing flight. Rocket motors would accelerate the craft to Mach 24, the velocity needed to enter earth orbit.

Eliminating the need for rocket power from the ground to around 200,000 feet would reduce launch costs enormously. And unlike rockets, a NASP vehicle would not have to carry enormous tanks of liquid oxygen to use as the oxidizer for its hydrogen fuel. The craft will need to carry only enough on-board oxygen to fire the engines briefly for a final "kick" into orbit, to maneuver in space, and eventually to re-enter the atmosphere.

A NASP-derived vehicle would compare favorably with other space launch technologies. According to a May 1990 study by OTA, the cost to launch an aerospace plane would run between $800,000 and $4.4 million per flight, for a payload of 20,000 to 30,000 pounds. That's far more economical than the shuttle, with its nominal launch cost of $275 million.

The NASP's cost advantages stem primarily from its similarities to ordinary aircraft. Operating out of an airport and taking off from airport runways roughly the size used by commercial airliners, NASP vehicles could be launched frequently. According to plans developed by the NASP program office, a NASP vehicle could typically take off 24 to 36 hours after landing. Where rockets often sit in hangars and on the launching pad for weeks or months while the payload is loaded and all systems checked, NASP payloads and fuel could be loaded just a few hours before takeoff. A NASP vehicle will require inspection similar to that given an airplane—not the intense scrutiny required by a rocket. (If NASP suffers a mishap after takeoff, it can simply turn around and land again; the shuttle and other rockets risk total destruction.) And NASP will not be much more sensitive to weather conditions than airliners are. By contrast, the shuttle often waits days for acceptable launch conditions. Finally, while safety considerations usually force rockets to be launched over oceans, NASP could fly over land. This flexibility gives NASP the advantage of access to virtually all low-earth orbits.

NASP relies on two critical technologies: an engine that can achieve hypersonic flight with little or no rocket assist, and structural materials that withstand high stress and high temperatures. Recent developments in both areas are encouraging.

When NASP was first proposed as a national program in 1985, there was widespread skepticism that high-performance materials could be developed to hold up to the extreme operating conditions. This lack of suitable materials has essentially been solved. Early in the NASP program, the major contractors—General Dynamics, Rocketdyne, Pratt & Whitney, McDonnell-Douglas, and Rockwell International—formed a consortium devoted to materials reached for the NASP. Each contractor is responsible for a particular materials class and then sharing the results with the others.

Over three years, the five companies have spent about $160 million, and their efforts have produced an array of materials that appear to withstand any anticipated stress. For example, structural elements in the air inlets or exhaust nozzles and near the leading edges will have to withstand temperatures of up to 3,000 degrees Fahrenheit. Traditionally, in other systems, those surfaces are faced with carbon-carbon (C-C) composites (carbon fibers embedded in a carbon matrix). At high temperatures, however, the C-C material oxidizes and erodes. This is why shuttle materials have to be replaced routinely.

The solution—considered highly improbable just two years ago—is a self-healing, glasslike coating for the C-C's surface. This coating prevents oxidation and withstands extreme changes in temperature. The design goal is for this material requirement to survive 150 flights without needing refurbishing. It has already withstood more than 250 flight demonstration cycles. General Dynamics is now making four-by-ten-foot structural panels of the C-C material. These panels are one-third the weight of conventional titanium structures, yet they withstand three times the temperature.

For structural components that don't experience quite such extreme temperatures, a team led by Rockwell International has developed a new form of titanium-aluminide (Ti-Al). Ordinarily, Ti-Al tends to be brittle when rolled into flat sheets. This new form, strong up to 1,300 degrees Fahrenheit, is produced by a new hot-rolling technique that assures uniformity of strength throughout the entire sheet. The sheets are now being fabricated into large structures using a diffusion bonding method that structurally links pieces of metal at the molecular level. Diffusion bonding eliminates the need for welding, which can weaken the material.

For critical areas where the enhanced Ti-Al is not strong enough, Rockwell has developed another version of the alloy that is reinforced by embedded silicon carbide fibers. By controlling the orientation of these fibers within the metal matrix, engineers can assure that the material's strength matches the expected loading on the part. It's like reinforced concrete, in which steel rods carry the load. Ordinarily, different expansion properties of the materials cause failure at very high temperatures, but that problem has been solved by high-purity processing. Contractors are also now producing four-by-ten-foot structures of this reinforced-titanium-aluminum material, including wing cross-sections that can withstand temperatures up to 1,500 degrees F.

The payoffs of this materials engineering will extend well beyond NASP. In particular, the ability to design a material's properties to match the stress it must bear is likely to significantly influence product design and manufacturing. Ten years from now, manufacturers may routinely design three-dimensional materials properties into their products thanks to the innovations of NASP developers.

Most essential to the NASP concept is its engine technology. The optimistic design objective is for these engines to propel the aircraft nearly all the way to orbit without rocket assist. So far that optimism is unshaken.

NASP's innovative propulsion system is a variation on a well-developed engine type called a ramjet. The entire front of a NASP vehicle would function as a large air collector, and the geometry of the structure would form the compressor. The massive amounts of machinery used in standard turbojets to compress the airflow would simply not be required. When the vehicle reaches hypersonic speeds, the engine will function as a "scramjet" (supersonic combustion ramjet).

Although the NASP test aircraft (designated the X-30) will not fly for several years, ground tests of the engine, as well as supercomputer simulations, have been encouraging. Engine efficiencies have reached about 70-80 percent, closing in on the goal of 90 percent—comparable to a typical turbojet. Nearly 50 wind-tunnel tests of scaled versions of the engine have simulated speeds of up to Mach 8, and there is strong confidence in the engine performance up to about Mach 10 or 12. Shock-tunnel tests well under way permit some analyses up to Mach 22.

As a manned, reusable vehicle, NASP will have the advantage of repetitive flight testing. These tests will allow designers to refine the operating characteristics and gauge the vehicle's performance, step by step, as it begins to probe unprecedented altitudes and speeds. And unlike a rocket, the NASP can have varying degrees of success. A rocket that does not achieve orbital velocity is a dud. But even if NASP fails to go beyond Mach 15 or Mach 20 in air-breathing flight, it would still be very nearly in orbit. It would simply need to carry more oxygen than the ideal to get its final boost into space. (Additional acceleration at high altitudes requires relatively little thrust since there is virtually no air resistance.)

NASP has had a difficult time finding its niche in the government bureaucracy. Facing massive budgetary pressures, Secretary of Defense Richard Cheney recommended canceling the program in April 1989. Vigorous objections from within the Defense Department and NASA, plus a favorable review by the National Space Council, stayed the axe, but funding has been cut significantly. Congress appropriated only $254 million for the program in fiscal 1990, down from the $427 million that would have been required to meet the original timetable. As a result, flight testing originally slated for 1994 will not now begin until 1997, delaying the availability of an operational aerospace plane until early in the next century.

The NASP program competes with familiar technologies, both in space access and in high-speed flight, that have long-standing constituencies. Many in NASA, for example, still view NASP as a competitor to the shuttle and its shuttle-like follow-ons. But NASA is becoming a more enthusiastic participant in the program. At the same time, many senior Air Force officials have embraced the program, although budgetary pressures have often appeared to place NASP on the auction block.

As the NASP gets closer to flight testing and produces operational results, resistance should diminish. The NASP program is now developing technologies—most visibly in the area of high-strength materials and reusable engines—that will benefit almost any new approach to space access, including mainstream efforts such as the Advanced Manned Launch System. This spinoff potential will doubtless build stronger support for the NASP program.

But it would border on tragedy if parochial turf wars blocked NASP from progressing at least to the test flight stage. Automobiles made their full impact only when driving a car became more than a hobby practiced by eccentrics. Computers have revolutionized business and society to the extent that they have become inexpensive, readily available, and easy to use. And only when travel into orbit ceases to be a newsworthy event can we claim to have truly entered the Space Age.

GETTING A LIFT TO LEO[3]

In the 1980s it seemed like almost everyone wanted to fly in space. Perhaps it's due to the glorification of America's astronauts still riding high in the media-mind as heroes and role mod-

[3]Article by John A. Dennison. *Ad Astra*. P 8+. Ja '90. Copyright 1990 by the National Space Society. Reprinted with permission of author.

els or maybe it's the popularization of space by institutions like Huntsville's Space Camp, which give Americans all the hype—mixed with hypervelocity—they could ever want about the joy and wonderment of spaceflight.

In reality, though, our society is a long way from launching tourists into space. To date, only the Space Expeditions Company of Seattle, Washington, has carried out a seemingly serious effort during the past decade to propose launching tourists into low Earth orbit (LEO).

That company originally proposed a launch date of 1992 to coincide with the 500th anniversary of Christopher Columbus' discovery of the New World. Space Expeditions' President T.C. Swartz recently refunded the $5,000 deposits of the 186 people who reserved rides into orbit, adding that "Since we cannot give you an exact target date and since it is so far behind our original schedule, we no longer feel comfortable in having your deposit in escrow." Riders were prepared to pay $50,000 for their lift to LEO.

Although Swartz has stated that space tourism is a "realistic concept," it is clearly not a concept which is ready to be implemented today or in the near future. The reasons for this are not purely technical. You can point to the fact that private launch companies are experiencing trouble similar to those that hinder space tourism—how to assess the real demand for the service, how to cut through government red tape and how to use technology to facilitate the service.

A 1986 Gallup Poll asked Americans if they'd like to ride on the Shuttle sometime in the future. A national average of 38 percent said yes. Of those surveyed, 50 percent of the men and 26 percent of the women responded positively. And this poll was taken *after* the *Challenger* accident of January 1986.

NASA, the United States' only source of piloted launch services, has always been the focus of public attention. The public, those footing the bill for all this spectacle, often feel they should have a chance to participate in the excitement of space travel too. Although most people realize that NASA's not in business to fly tourist-class astronauts, the wish to fly was so strong that many wrote to the space agency to express their desire.

Former NASA Administrator James Beggs recalls receiving substantial amounts of mail from would-be space tourists: "There were letters from Bob Hope, John Denver and Jacques Cousteau.

There were neatly typed resumes and hand-scrawled notes. There were elegantly embossed envelopes from the rich, and colorful, laboriously hand-decorated manila envelopes from the young. Thousands and thousands [of people] wanted to fly in space."

With this seemingly incredible demand for such a service why is it that the United States, a land of consumers, can't find a provider for the service of space tourism.

The fundamental problem is one of economics. The Shuttle, by some calculations, costs an estimated $500 million to launch. This figure divided by five, the size of the average crew, means it costs $100 million person to fly in space. Clearly, these costs will have to decrease by at least a factor of 100 before even Donald Trump could afford a Shuttle trip other than from the fleet he already owns.

The expense of developing and building the first three operational Shuttle orbiters has been estimated at roughly $40 billion. If this cost could be reduced by a factor of 10 and the payload capacity of the proposed vehicle increased tenfold, then costs may be in line with what it might take to develop and build a spacecraft for tourists.

Yet, once the considerable economic barrier is overcome, governmental and legal hurdles abound. All the primary launch sites in the United States are government-owned and companies using these facilities must follow government rules and provisions to use them. In addition, like other entrepreneurial efforts, new, untested vehicles have problems obtaining launch insurance. Also, it is expensive to modify or rebuild existing facilities for new spacecraft.

T.C. Swartz's Space Expeditions Company faced many of these types of problems. "The amounts needed for space exploration must be supported by government infrastructure, as the financial numbers from a purely commercial aspect become unrealistic without this kind of infrastructure assistance," he noted. Swartz decided to seek private funding through venture capital groups and looked into a public stock offering. Prospects appeared good until the stock market crash in October 1987 curtailed risk-taking investments, particularly those of the scale needed to support a space tourism enterprise.

Should NASA help start-up companies that want to pursue space tourism?

Space endeavors with important scientific and medical purposes have been given assistance, such as free Shuttle flights, by NASA under a program of Joint Endeavor Agreements (JEAs). Unfortunately, separate funding from NASA's budget has not been provided for these JEAs. Since NASA must therefore remove money from one of its own branches to provide this support, no department in NASA is thrilled about having their resources used on someone else's project. Many companies, running on venture capital and having only a short time to establish themselves and turn a profit, have established JEAs and are then delayed by NASA for a year or more—by which time the companies may have folded. It is unlikely that space tourism would fare well under this program since other projects, supposedly of more national importance, have not.

Tourism would also conflict with the exclusivity of NASA's ability to put humans in orbit, threatening the agency's prestige and image. How would it look if anyone could fly in space at a reasonable cost instead of a tiny minority of highly-educated and trained astronauts that made it through NASA's selection process?

There are two major drivers, either of which must happen before space tourism can begin. The first is that the private sector must be in control of launching space vehicles or at least be able to do so on a routine and affordable basis.

The second possibility is that another spacefaring nation, such as the Soviet Union, or a spacefaring wanna-be, such as Japan, began to offer tourism services first. In fact, this has already begun. The Soviets and Japanese have already entered a partnership to fly a Japanese journalist to MIR this decade [In December 1990, a Japanese journalist went into space and visited the Soviet space station MIR. His employer paid the Soviets 12 million dollars for the privilege.] The Soviets are also training two MIR-bound private citizens from the United Kingdom, the fare to be paid by a British advertising agency. A Soviet journalist flight program is also now in the final planning stages. The U.S. government may decide to support made-in-the-USA tourism efforts just to save face.

Seattle's Space Expeditions Company has been the only serious effort to offer rides to LEO, but it hasn't been the only tourism proposal. The Space Trust Corporation, a Florida-based operation, has proposed a novel approach to putting folks in orbit: a 55-seat passenger accommodation compartment that would fit into the Shuttle's cargo bay. Space Trust also proposed signing up 2.5 million members at $50 a head for a lottery to see who'd actually get a Shuttle seat. The group would also plan to solicit corporate sponsorship so the lucky lottery winners would have to pay few, if any, additional fees. NASA has not rejected the idea outright, but it seems likely that the idea will remain in an indefinite bureaucratic holding pattern.

One new effort that may lead to space tourism is a project called Space Phoenix based in McLean, Virginia. The group is consulting with experienced members of the space community, such as former NASA Administrator James Fletcher, about converting one of the Shuttle's external tanks (ETs) to an orbiting laboratory.

In theory, astronauts could rework the 70,000 cubic feet of space designed to hold liquid hydrogen and liquid oxygen into pressurized, habitable volume that could house scientist and engineer "tourists" and their experiments. Space Phoenix planners estimate the conversion costs to be only a few hundred million dollars—a veritable orbital bargain.

Once again, government cooperation is the key. Space Phoenix needs to secure permission from NASA to take possession of a Shuttle ET and arrange to have it deposited in orbit by a Shuttle. Then the Shuttle would be required to ferry the personnel needed to convert the ET to a habitable laboratory. It remains matter of speculation as to when the public could visit—and to the mode of transportation used to get them there.

One option for those "eagernauts" among us to experience space now, but without benefit of orbital flight, is offered by the American Experimental Spaceflight Association in Carteret, New Jersey. The group's invitation, "To Fly to the Edge of Space," discusses the space-like views possible on a chartered stratospheric supersonic flight. For $25 a year you get a newsletter and a chance to ante up for the chartered Concorde flight.

While space tourism is still a fantasy for now, someday, maybe sooner than anyone expects, it will become reality. The few organizations which have opened memberships to the public or offered opportunities to fly have met with great response and seem to represent only a fraction of those who'd like to experience space travel. Shuttle passenger Senator Jake Garn (R-Utah), one of the first "tourists" in orbit put it this way: "The people that go will have the most fun they've ever had in their lives."

VOYAGE TO A FAR PLANET[4]

Aboard the spaceship *Glasnost*, 23 million miles on its way to Mars, Baryshnikov (so dubbed for his annoying habit of repeating the *Swan Lake* adagio for hours in the low-g exercise module) had stopped shaving his face. He had elected to use his electricity ration to do his armpits once a week instead. Spiderman was indifferent to this. He was more interested in the web of Kevlar monofilament he'd woven in the radiation shelter. Besides, he hadn't changed out of his black tights for half a year and Baryshnikov was the only one who had said nothing—so a silent bargain had been struck. Baryshnikov, a.k.a. F. Povich, cosmonaut of the USSR, and Spiderman, a.k.a. P. Hoffman, astronaut from Houston, had shown the unmistakable signs of irreversible space psychosis.

The six other crewmen had their f-a-c-u-l-t-i-e-s somewhat more intact. They had gathered in the galley to make the decision.

"They could sabotage the entire ship," Akira from Nagano, Japan, had pointed out.

"We must not act in an uncivilized and immoral fashion, however," advised Mohan from Bombay.

"We must conform to due process of law," Jean-Paul the Parisian agreed.

They decided to isolate Baryshnikov and Spiderman in the exercise and shelter modules. That way, if they blew an airlock, started an electrical fire, or found a more novel method to merge their souls with the black void, they wouldn't take the other six astronauts along with them.

The six did not discuss their decision with ground control. After all, who knew what the others back at the lunar base, or at mission control, might do?

The six were having a little trouble with the isolation, too.

[4]Article by Brenda Forman. Reprinted by permission of *Omni*, © 1990 Omni Productions International, Ltd.

"A trip to Mars will be the most challenging high-technology adventure of the twenty-first century," says Tom Paine, former NASA administrator and chairman of President Reagan's National Commission on Space. Getting there, however, makes a lunar flight look like a jaunt to the corner store. While the moon is a "mere" 250,000 miles away, Mars can be anywhere from 49 million to 235 million miles from Earth, depending on the time of year. One round-trip could take three years or more, not counting time spent on the surface.

Along the way, human colonists will live for months or years in cramped spaces. In the meantime, politicos will be maneuvering back on Earth, persuading half a dozen nations to split the tab on this pricey journey. The $400 billion (NASA's preliminary estimate) sticker shock, however, isn't the only reason that the United States is unlikely to go to the red planet on its own: The space technology that such a big journey demands is no longer exclusively Made in America.

The Soviet Union, the European Space Agency (ESA), Canada, Japan, and China all have very impressive space capabilities—including rockets, robotics, and space-medicine expertise. Recognizing this, President Bush announced in March that the United States plans to seek an "exploratory dialogue" with Europe, Canada, Japan, the Soviet Union, and other nations on international cooperation in the Space Exploration Initiative.

Even though no one knows what shape such "international cooperation" will ultimately take, there's certainly enough scientific and technological work to involve everybody.

"Every way you look at SEI [Space Exploration Initiative, i.e., moon-Mars in government speak,] it's a whole cluster of cooperative opportunities," says Peter G. Smith, director of international relations for NASA. Fitting the pieces together will be the most complex systems integration job ever. Meanwhile the project's international organizational structure will have to function for 30 years or more—a staggering prospect given the routine mulishness of politicians and the vagaries of history. (Remember the furor last year when NASA offered Congress a scaled-down space station design without first consulting the other nations involved with *Freedom?*) Even before the first crew leaves for Mars, there's a formidable clutch of technologies to master—both "hard" ones for hardware and systems, and "soft" ones for keeping space colonists sane and healthy.

"The Americans are talking about cutting their Mars mission budget once again, sir."

"They do this to us every year, don't they? You'd think they'd persevere in something they started themselves, but they never do. I suppose we should have learned that back in Space Station *Freedom* days."

"Minister says we should think about redesigning our module to be completely independent. Then we can let the Americans dither without queering our own projects."

"Mmm. More expensive that way, of course. But worth looking at."

The politics of moon–Mars will probably be absolutely Byzantine. "The technological challenge is probably equaled by the institutional challenge: to organize a sustained international effort extending over decades," explains Paine. Sending voyagers to Mars will involve a broad consortium of nations and stretch over three decades or more.

The only project remotely resembling such an ambitious endeavor, Space Station *Freedom*, has the United States as its leader, manager, and ultimate arbiter of operating decisions. "That could be the Mars model—if the United States cares to pay the bulk of the gigantic amount of money involved, as it is doing with the space station," says Ben Huberman, President Carter's deputy science adviser and now a Washington consultant on technology issues. Even more important, the nation would have to commit to paying out that money consistently over the extended time period of a Mars mission, a condition Congress will undoubtedly reject.

The *Freedom* model may not prove flexible enough for the moon-Mars initiative. Although big and complex, the space station is essentially just one single piece of hardware. Moon-Mars will require a very large number of separate elements—bases, stations, launchers, and landers—each of which could involve different players. For example, according to a high NASA official who requested anonymity, Japan is already deeply interested in lunar resources, so it might want to participate in a moon base. The Soviets, on the other hand, prefer a Mars mission over a return to the moon. Each piece of the action, therefore, will probably end up tailored to the various participants' goals.

Paine says a good management model might be Intelsat (International Telecommunications Satellite), to which each nation contributes to the degree it uses the system: "Pay your dues and get your ride." Another model: the approach ESA has pioneered in coordinating the space efforts of 13 countries with different technology levels, funding resources, and policy objectives.

But would the United States agree to any arrangement in which it would be a mere partner rather than the leader? "I don't think we'll go to Mars as part of an organization in which we lose the perception of the lead," says the same anonymous NASA official. "That tugs at the U.S. heartstrings. Leadership is one of the main reasons the President put forward the program."

Such nationalistic fervor could prevent Earthlings from ever reaching our red neighbor. Should the world's spacefaring nations overcome their instincts to go it alone, however, here's what a joint effort might lead to in the next 40 years.

"Ten, nine, eight . . . " The mammoth rocket sits on the Baikonur launchpad deep in the south central region of the Soviet Union, poised to truck another 100 metric tons of hardware to low Earth orbit. This is the twentieth such launch, and astronauts and cosmonauts in orbit are busily assembling all the payloads into a convoy headed for the moon. The booster is the Energia, pride of the Soviet launch stable and the biggest vehicle to rise from a pad since the Saturn 5 of Apollo days. Indeed the two rockets are in the same class: Energia can lift 220,000 pounds to low Earth orbit; the Saturn 5 lofted the 165,000-pound *Skylab*.

The first step to Mars: propelling thousands of tons of construction materials, rocket motors, spacecraft components, life-support systems, living modules, air, water, food, and humanity out of Earth's gravity well and into space. Neither the shuttle, with its 48,000-pound capacity, nor the biggest U.S. rocket, the Titan 4, which carries 40,000 pounds to orbit, could deliver such a massive amount of equipment. And what will be ESA's largest launcher, the Ariane 5, will lift only 46,200 pounds.

Comparing the Mars mission's lift requirements with those of the space station gives some feel for the magnitude of the effort. At present there are more than 380 station assembly elements scheduled for delivery to low Earth orbit on about 30 shuttle flights over a period of four and a half years, scheduled to begin in the mid-Nineties. Such an orbiting platform would be just one part of a moon-Mars program. A lunar base itself would need far more launch capacity than the station. Besides the living modules, rockets would also have to hoist heavier power plants, surface rovers, and more complex communications gear.

Only the Energia has the capacity to boost such tonnage to space. "The Soviets have built a heavy cargo vehicle," says Art Dula, president of Space Commerce Corporation, which is marketing Soviet launch services in the United States. "We need a dump truck and all we have is the family Rolls!"

Mere technical grounds rarely win this sort of argument, however, so don't expect the United States to depend on the Energia for its heavy-lift needs. It would be more likely to revive work on the Advanced Launch System (ALS). ALS started in 1987 as a program to revolutionize American rocket technology and lower the cost to orbit by a factor of ten. Lately the effort has had its funding cut for lack of a clearly defined mission. Moon-Mars just might turn out to be that mission.

Ultimately, nuclear fission or something even more advanced—such as fusion—will probably be the only way to bring down the cost and one-way travel time. Nuclear propulsion might cut the trip to 100 days from the 550 days a chemical rocket would take. Although Westinghouse started working with Los Alamos laboratories on this approach in the late Sixties, according to an industry insider who requested anonymity, the technology is still not operational. To get the voyage down to 50 days would take even more advanced motors—such as those using helium 3 for nuclear fusion. Unfortunately, after millions of dollars of government-funded research, scientists are still unable to control fusion reactions.

NASA's Lunar Energy Enterprise Case Study Task Force recently issued a report on helium 3, which is rare on Earth but plentiful on the moon. The group suggested that in the future the element could fuel fusion reactors to power spacecraft.

"Mass approaching twelve o'clock, diameter ten centimeters, speed six kilometers per second, impact imminent!" The computerized voice echoes through the spacecraft; the impact shakes the structure. Everything necessary to repair the ship must be either available onboard or jury-rigged by an ingenious crew. There is no alternative: The voyagers are 3 million miles from Earth, and Mars is another 45 million miles ahead.

"You can't have a Cape Kennedy in orbit," declares Mark Craig, NASA's special assistant for exploration and point man for the moon-Mars initiative. "So how do you build a spacecraft that will exist in space for years with little or no human intervention? You need self-diagnostic ability and repairability."

Current manned space vehicles have an average MTBF (means time between failures) of 10,000 hours; in other words, more than half of them may fail within a year. Even on the Soviet *Mir* space station, where replacement parts can arrive within 30 days, the crew spends large amounts of time repairing equipment, observes Nicholas L. Johnson, advisory scientist at Teledyne Brown Engineering and author of the annual compendium The *Soviet Year* in Space. "The Soviets have had significant problems, especially after the first two to three years, in maintaining the *Mir*," he says. "They've done a great job, but it's been at the cost of a lot of lost cosmonaut time for conducting experiments." That's in low Earth orbit. For any deep-space mission, repairability is essential. "If you can't fix something," says Johnson, "you'd better have another one along with you."

The robotic servicing unit's TV optical system transmits the faint glow of the ruptured nuclear power plant to the operator bent over her screen three kilometers away in the lunar module. *Damn*, she thinks, *yesterday's meteor impact breached the first two layers of the triple shielding; any more damage would have caused a meltdown.*

No human could go near a reactor in this condition and survive. Closeup repairs of the shielding are no problem for her teleoperated robot, however. Its "eyes" let her zoom in on the damage; the "smart" gloves she wears let her move the robot's "hands" to feel the reactor wall. Thanks to the dexterity with which she can manipulate her distant partner, repairs are finished in a single eight-hour shift.

Making repairs outside a ship in space is a risky business for humans. When possible, spacefarers will employ futuristic teleoperated or fully independent robots equipped with sensors able to distinguish texture, "hands" with "fingers" that can feel delicate gradations in pressure, plus enough artificial intelligence to learn from experience and program themselves. "This is an area where Japan can contribute a lot," says Hajime Furuta, MITI representative in New York, referring to Japan's demonstrated virtuosity in robotics.

Canada, the nation that gave the shuttle its manipulator arm, is also a major player in robotics and automation. At present Canada-based Spar Aerospace is developing the Mobile Servicing System that astronauts will use to build and maintain the space station. Furthermore, robotics will play a major role in mining the lunar regolith for construction materials and rocket fuel, and in covering habitation modules with soil to protect against radiation. "We have an active mining community," says Dr. Frank Vigneron, who chairs the Canadian Space Agency's Working Group on Moon-Mars Exploration, "so there's possibly a role for us in the production of robotic vehicles to do similar work on the moon."

Mining, construction, and other robotic activities, however, will consume more energy than existing power-generation techniques can supply Space Station *Freedom's* big solar arrays, for example, are expected to provide a total of about 75 kilowatts—barely enough to keep the station running and do any work onboard.

A moon-Mars mission will require far more power than the station. "I think nuclear power will be needed," says Huberman. "The United States is working on the SP-100 nuclear power source, and the Soviets run their Rorsat with nuclear reactors. So to the extent that nuclear is required, it will be an American-Soviet show."

"Two dozen suits in the locker and not a damn one working in my size!" fumes the geologist intent on getting out onto the Martian surface while the pink Martian sky is still bright with sunlight. Within limits, the suits can be resized to fit different human frames, but right now all the ones suitable to her height and build are out of commission. Whether in the vacuum of space, the airless surface of the moon, or the tenuous atmosphere of Mars, the fragile human body must be sheathed and supported by the elaborate protection and life-support systems of a space suit.

The International Latex Corporation space suit used on the shuttle is a remarkable piece of gear—but is too delicate for regular wear in space or on the dirty surfaces of the moon or Mars. The garment needs hundreds of hours of repair and reconditioning after every spacewalk and takes days to resize to a differently proportioned person. It also takes hours to get into one and prepare to leave the ship.

To avoid developing the bends, even before dressing, the intrepid traveler must spend from 40 minutes to four hours breathing oxygen before setting out on the EVA. The reason: Shuttle space suits operate at a very low 4.3 pounds per square inch (psi).

First an astronaut dons a liquid cooling and ventilation garment, designed to keep internal temperature bearable, laced with a network of flexible tubing. At this time the urine collection device is also put on. Next come the lower torso pants, boots, hip, knee, and ankle joints. Then follows the upper torso, including arms, the umbilical and electrical harness containing communications, power and oxygen lines, connections to the portable life-support systems—and the Hamilton Standard life-support backpack. Finally the spacefarer slips into gloves, helmet, and visors.

Unfortunately, shuttle suits are too heavy for use in the lunar or Martian gravity fields. Built for zero gravity, one weighs more than 200 pounds on Earth. "The weight has to come down," says Lee Weaver, a California consulting pilot engineer who has worked in every EVA program (and worn these suits) since Gemini. "The suit has to have better lower body mobility and it has to be able to stand up to the lunar or Martian environment.

"EVA is going to be a make-or-break factor in assembly and surface operations," he continues. "Unless and until we have nuclear power to support planetary surface operations, human muscle is going to be one of the primary ways to make things happen." That will require major advances in space suit design.

NASA is now working on better garments. At Johnson Space Center a team is developing an 8.6 psi outfit that would eliminate the need to breathe oxygen for hours before an EVA. Unfortunately, the suit has to be sturdier to handle the increased pressure. "That, in turn, makes glove dexterity a problem," says Weaver. "And you can't use pure oxygen at that pressure because it's too flammable, so you have to use both nitrogen and oxygen, which requires the addition of pumps, regulators, and gas sensors. That adds to the weight—and we need lighter, not heavier, suits to operate in the gravity fields of the moon or Mars."

A completely hard suit, the AX5, is in the works at NASA's Ames Research Center. Its all-aluminum shell offers a vast safety improvement over soft suits, according to Weaver. It does have a challenge to meet, however: In a gravity environment the AX5 will have to be modified to stand up by itself when inflated (as a soft suit will).

"Ahhh, that's better! Gravity at last. Ten years in the space biz and I still get spacesick when we hit orbit!" The computer systems troubleshooter relaxes gratefully onto her couch and lets her stomach settle down as the habitat module starts to spin at the end of its kilometer-long Kevlar tether and the resulting centrifugal force creates a modest level of gravity. Tomorrow an elevator will crawl up that cable, taking her to weightlessness again at the spin center of gravity, where she will attempt to coax a balky computer system into functioning properly. For now, at least, she'll be able to relax in comfort—and keep her dinner down.

The absence of gravity does very nasty things to the human body. After even a few days, astronauts lose bone calcium, cardiovascular conditioning, and electrolytes. The most commonly used preventive measure discovered to date: two to four hours of strenuous exercise every day. On a three-year mission such a workout schedule might prove both hard to maintain and inadequate to the cumulative health impact of zero g. Some form of artificial gravity, therefore, may play an essential role in the Mars voyage.

One way to create gravity is through the centrifugal force generated by spinning two objects (such as a spacecraft and a counterweight) connected by a long "string."

Such tethers will be long cables of woven Kevlar deployed like huge fishing lines from space platforms. Research into this technology has progressed quietly in recent years—to the point where the shuttle is slated to test one in September 1991. Plans call for astronauts to uncoil a 20-kilometer-long tether with an Italian satellite at its end. How much time might it take to scale up from this demonstration level to an artificial gravity system for a functioning spacecraft?

"Not too long," predicts Al Schallenmuller from Martin Marietta, which has been working with NASA for years on tethers. "The main areas that we need to master are some of the dynamics in the cable and some of the electrostatic charges generated in it when it's flying in space."

Another way to simulate gravity: Create centrifugal force by spinning the spacecraft alone. "But if you do that," says Larry Bell, director of the Sasakawa International Center for Space Architecture at the University of Houston, "you also create problems with orienting communications antennas, radiators, and solar power arrays." Beyond these difficulties, the craft must rotate in a wide enough arc to avoid causing an inner ear disorder called the Coriolis effect, which results in a loss of balance. "With a radius of fifty-six feet, we can reach six revolutions per minute, which provides two thirds the earth's gravity," says Schallenmuller. "But higher rotation rates," he adds, "would require a longer radius and create a need to shore up the structure to brace it against propulsion stresses."

"How many more flights are they going to allow you?"

"Only one. I'm already pushing the radiation dosage limit."

"That's a heartbreaker. Who's going to finish out your experimental series?"

"One of my graduate students, I think. He's smart enough— if he decides he wants to risk the possible chromosome damage. I think he wants kids eventually. He just doesn't want them to have two heads!"

The two scientists laugh hollowly, contemplating the difficulties of conducting research in space. After all, some necessary experiments take longer than human tissue can withstand the onslaught of solar and cosmic radiation.

Experts in space medicine now place such risks at the top of the health hazards list. "We're using standards that are very, very speculative," warns Bell. "We're already allowing dosages for astronauts that are ten times what we allow for radiation workers on Earth. It's not simply a matter of adding shielding, either. That increases the ship's weight, and far worse, spacecraft walls can become ionized, making the radiation problem worse. This issue is a potential showstopper—and it's the one we know the least about!"

NASA is now talking to the Soviets about their space medicine experiences. The agency has even discussed joint research efforts—such as flying U.S. astronauts and using American medical protocols aboard *Mir*. Two important areas in which such a program would help: an assessment of the need for artificial gravity and research on radiation effects.

"Well, let's see, you raise the fish in the rice paddy tanks, and you filter the water by running it through the soil in your vegetable garden—you better have enough termites in there to clean up the dirt, by the way—and oh, yes, I'd stock some chickens that still know how to sit their eggs—they bred it out of the stock twenty years ago, you know—and . . . " The space ecologist talks on, while his listener strives to grasp the complexity and expense of equipping a completely self-contained spaceship capable of taking his Martian Tours customers ("For the Traveler Who's Been EVERYwhere!") safely and comfortably to the red planet.

To stay alive in orbit and on the lunar and Martian surfaces, people will need closed life-support systems that can recycle air and water, grow food, and stay habitable for years or even decades at a time. There is nothing of the sort available now. Neither the shuttle nor *Mir* has closed life support; both depend on resupply from Earth. The shuttle lands between missions. Approximately every 60 days the Soviets launch a Progress to *Mir* carrying two and a half metric tons of various supplies—such as propellants, water, air, food, film, and space suit spare parts. "To support a two-man crew for a year takes five to six such flights," says Johnson. "That's about fifteen tons of material every year for just two people." Although Space Station *Freedom* originally had a closed oxygen system, last year that idea was dropped in favor of the *Mir* approach, to save money. No such option will be available for a Mars journey; there aren't any supply depots along the way. It's recycle or die.

Let's assume engineers master all these problems of equipment and hardware. The hardest part still remains: keeping fragile, ornery humans physically and psychologically healthy for months and years cooped up in spacecraft or surface structures the size of biggish school buses, in a hostile environment, away from family, friends, fresh air, blue skies, and the earth's gravity field. "The human being is the weak link in the chain, and it's a very weak link," says Susanne Churchill, associate director of the Institute for Circadian Physiology in Boston. "We typically put enormous effort and money into building spacecraft but nowhere near the necessary level of work into the matter of the crew. Unless we focus on the human aspects, we may engineer ourselves a perfectly elegant ship to launch to Mars—and find we can't in good conscience put people in it."

"I can't stand it any longer! If you whistle that tune one more time, I'll kill you!" The technician clings to a hand brace on the wall of the spacecraft's galley, menacing his shipmates with a knife, his eyes wild. The others hover at a cautious distance while the ship's psychologist tries to calm the distraught man down. It's his third crack-up this week. *The trip's too long; we're too far away from home*, she worries. Even the best prelaunch psychological profiles can't predict what will drive people crazy several million miles from Earth.

Spaceflight is tough duty even close to home in low Earth orbit. When this reporter first met Dr. Oleg Atkov, the Soviet cardiologist who spent nearly eight months in orbit on *Salyut 7*, she mistakenly complimented him on enduring 236 days here. "Two hundred thirty-seven," he ruefully corrected, "because every day you count!" With cosmonauts staying in orbit for as much as a year at a time, the Soviets have accumulated more experience than any other nation in the psychological problems of spaceflight. The techniques they've developed attest to the complexity of the problem. Every mission has a full-time ground-based psychological support group. Cameras onboard the space station allow psychologists on the ground to monitor crew interactions, watching for indications of tension. Cosmonauts' voices are monitored for signs of stress. Regular radio and TV contacts with families, friends, and prominent Soviet personalities boost crew morale. Mail, pictures, and tapes of Earth sounds go up on every Progress resupply module. Thus far, it's worked. Nobody has murdered anybody else in space. But that's only in low Earth orbit. Psychologists will need a whole new set of techniques to deal with people who are many millions of miles away for years at a time. Perhaps by the time people venture to Mars, advances in computer and other technologies will let them talk to Hal-like psychologists based in the ship's computers, gaze at holograms of their loved ones and their favorite vistas, or dream in some form of suspended animation to pass the time.

If we overcome all these technical and political hurdles, the day may come when, finally, humankind sets out for Mars. The journey might go something like this: Dateline 2030: Built by the biggest consortium of nations in history, piloted by a multinational crew, supported by an international lunar base, and fueled by propulsion materials mined from the lunar regolith, the huge spacecraft, assembled in lunar orbit, is finally poised to depart for Mars. Supplies are waiting for it there, pre-positioned by un-manned vehicles on the surface of the planet.

To have gotten this far is a triumph of technological wizardry. But even greater is the political triumph. Persevering somehow through four decades of national rivalries, budget crunches, and conflicting priorities, the participants have built cooperative in-stitutions and techniques for managing conflict that may look oddly like the nucleus of a world government. History may note that in going to Mars together, the world finally discovered how to work together on Earth.

IS SPACE THE PLACE TO BE?[5]

After a 32-month hiatus, the United States space program lurched forward last fall with the launch of the *Discovery*. Four days and a successful landing later, the flight was pronounced a nearly flawless return to space. With a redesigned Space Trans-portation System and an agency-wide shake-up, NASA is pro-claiming a new beginning to America's space exploration.

A new beginning it may be, but some critics of the space pro-gram find that NASA's underlying approach to space is un-changed. Many decry the environmental cost of space flight. Each shuttle launch creates about eight million pounds of toxic waste in the form of water tainted with hydrochloric acid, points out Jim Heaphy in the environmentalist magazine *Earth Island Journal* (Winter 1987). Many of the satellites launched with the shuttle continue to carry radioactive materials to power small nuclear re-actors or other electricity-producing devices. A number of acci-

[5]Commentary by James Marti. *Utne Reader*. P8+. Ja/F '89. Copyright 1988 by LENS Publishing Co. Reprinted with permission of author.

dents involving these nuclear payloads have occurred with both American and Soviet satellites; a particularly public example was the Soviet Cosmos 954, which in January 1978 crashed in the Canadian wilderness, spreading large quantities of radioactive material over thousands of square miles. *Technology Review* (Aug./Sept. 1988) also reports that the upper atmosphere has become littered with disabled satellites, spent boosters, waste ejected from spacecraft, and myriad bits of space junk.

Environmentalist Eugene C. Hargrove feels these incidents are symptomatic of a serious problem. In *Beyond Spaceship Earth: Environmental Ethics and the Solar System* Hargrove suggests that an examination of environmental ethics related to space exploration is long overdue. He poses questions that might soon become pressing: Should strip mining be permitted on the moon? Should space factories be allowed to discharge their effluent into the vacuum, on the grounds that no one can breathe it anyway? Is there anything wrong with restructuring Mars or Venus on a planetary scale (known to space buffs as terraforming) so that they might someday support Earth-like biospheres?

These long-term concerns are part of a larger appraisal of the whole philosophy of space travel—appraisals that for the most part are occurring outside NASA. Fundamental issues, such as the reasons for venturing into space at all and rules of conduct once we get there, have recently been addressed in the alternative press. In *New Age Journal* (Jan./Feb. 1988), Frank White compares exploring space to erecting the Great Pyramids and building the Gothic cathedrals of medieval Europe. All these efforts are seen as "central projects" for mankind, massive group efforts that use a material goal as a catalyst for social and personal transformation. White also likens space exploration to the first tentative forays of primeval fish onto land. The pioneer fish had no way of knowing what tremendous evolutionary opportunities they were opening up by venturing onto land; indeed, the fish species realized no gain from such a move. It was humans and other species that reaped the rewards of the fishes' land exploration program.

A less optimistic appraisal of the promise of space comes from Andrée Collard, in a posthumous article in the feminist journal *Woman of Power* (Spring 1988). Collard saw space exploration as incapable of providing any meaningful transformation of humanity: "Space exploration may bring . . . a better understanding of the solar system. It may even bring cooperation betweeen inimical nations. However, without a thorough overhaul of human values before we get there, the colonization of space will simply transport our 'planetary hangups' somewhere else."

A theme common among space program boosters is space as the "high frontier," a wide open place for human settlement and expansion. Some supporters of the high frontier concept argue that humanity has simply "outgrown" the limits of the Earth, and thus needs to expand into space. Implicit in such a view is the faith that this new frontier will act as a social and economic relief valve, much as the American West appeared to the urban societies of the 19th century. However, the analogy with past frontiers does not quite fit the present one. As William Hartmann notes in *Beyond Spaceship Earth*, "The Earth turns out to be a Hawaii in a solar system full of Siberias. . . . Unlike some early frontiersmen who exhausted one farmland and moved onto the next, we will find no rational motivation for destroying the planet to which we are umbilically linked and then attempting to move on." Critics of the space frontier theory argue that space exploration, for all its potential, cannot succeed in removing the limits to growth imposed by the Earth's environmental capacity.

The current thinking of space as a place for commercial exploitation and military installations is criticized by Gar Smith in *Earth Island Journal* (Winter 1987). "The great tragedy of the Star Wars debate is that it has overshadowed every other vision of humanity's role in space," he writes. "There is no longer much talk in the United States of designing solar energy satellites to provide the planet with a cheap, non-polluting source of power. Nor is there any talk of peaceful, international cooperation in space." Smith finds that among the many opportunities space presents, those favored by space policymakers tend to focus on the corporate and military approaches that have proved so unmanageable on Earth. "Before novice pilots are allowed to take off into the sky," he cautions, "they must first graduate from ground school, and from all evidence at hand, it would appear that our species has not even learned to taxi safely."

BIBLIOGRAPHY

An asterisk (*) preceding a reference indicates that the article or part of it has been reprinted in this book.

BOOKS AND PAMPHLETS

Alexander, Kent. Countdown to glory: NASA's trials and triumphs in space. Price Stern. '89.

Aldrin, Buzz and McConnell, Malcolm. Men from Earth: An Apollo astronaut's exciting account of America's space program. Bantam. '89.

Allen, Patricia. Among the heavens: The human side of spaceflight. HFSI Inc. '89.

Asimov, Isaac. The world's space program. Gareth Stevens.'90.

Bova, Ben. Welcome to moon base. Ballantine. '87.

Byerly, Radford, Jr. Space policy reconsidered. Westview. '89.

Collins, Michael. Carrying fire: An astronaut's journey. Bantam. '89.

Cooper, Henry S., Jr. Before lift-off: The making of a space shuttle crew. Johns Hopkins. '87.

Feynman, Richard. What do you care what other people think? Further adventures of a curious character. Bantam. '89.

Gump, David P. Space enterprise: Beyond NASA. Greenwood. '89.

Hofman, Helenmarie. Starlab! The NASA guide to student projects in space. Woodbine House. '91.

Kirby, Stephen & Robson, Gordon, eds. The militarization of space. Lynne Riener. '88.

MacKinnon, Douglas and Baldanzia, Joseph. Footprints: The twelve men who walked on the moon reflect on their flights, their lives & the future. Acropolis. '89.

Makower, Joel. The air and space catalog: The complete sourcebook to everything in the universe. Random House. '90.

McDougall, Walter A. The heavens and the earth: A political history of the space age. Basic Books. '85.

Michaud, Michael A. Reaching for the high frontier: The American pro-space movement. Greenwood. '86.

Miles, Frank and Booth, Nicholas. Race to Mars: The Harper & Row Mars flight atlas. Harper & Row. '88.

National Commission on Space. Pioneering the space frontier. Bantam. '86.

National Research Council. Mission to planet Earth. National Academy Press. '88.

Nelson, Bill. Mission: An American congressman's voyage in space. Harcourt Brace Jovanovich. '88.

Oberg, James. Red star in orbit. Random House. '81.

Ride, Sally K. Leadership & America's future in space. US General Publications Office. '87.

Smolders, Piet. Living in space: A manual for space travelers. TAB Books. '86.

Taubenfeld, Howard. Space and society. Oceana. '64.

Trento, Joseph J. Prescription for disaster: From the glory of Apollo to the betrayal of the shuttle. Crown. '87.

Wilford, John Noble. Mars beckons: The mysteries, the challenges and the expectations of our next great adventure in space. Knopf. '90.

Wolfe, Tom. The right stuff. Farrar, Strauss & Giroux. '83.

ADDITIONAL PERIODICAL ARTICLES WITH ABSTRACTS

For those who wish to read more widely on the subject of Our Future in Space, this section contains abstracts of additional articles that bear on the topic. Readers who require a comprehensive list of materials are advised to consult the *Readers' Guide to Periodical Literature* and other Wilson indexes.

APOLLO AND CHALLENGER

Lunar Reflections. Jeff Goldberg *Omni* 11:345+ Jl '89.

Part of a special issue, celebrating the 20th anniversary of the first moon landing. Sixteen *Apollo* astronauts share their memories and reflect on space travel. A sidebar describes what some of the astronauts carried in their pilot preference kits while in space.

Holocaust in the sky. *Discover* 7:38–51+ Ap '86.

A special section examines the *Challenger* disaster. The accident was a blow to NASA's reputation for technical excellence. Preliminary speculations focus on budget and performance pressures that may have pushed the program beyond its limits, threatening flight safety. Booster failure was only one of several catastrophes waiting to happen. NASA's apparent awareness of safety problems raises the question of why the program continued to accelerate.

The astronauts after Challenger. James Reston Jr. *The New York Times Magazine* 46–7+ Ja 25 '87.

The *Challenger* disaster a year ago has forced many of the nation's astronauts to come to terms with the risk inherent in space travel. Since the accident, 18 astronauts have either left NASA or been assigned to management posts. Those who remain on active flight duty have seen their scheduled flights put on hold, and some have found themselves in the position of saying I don't want to go, a statement that is an anathema to a test pilot. Astronauts Capt. Frederick H. Hauck, Anna L. Fisher, Comdr. Dale A. Gardner, Joseph P. Allen, and David M. Walker discuss the effects the Challenger explosion has had on them, on fellow astronauts and on the US space program.

The heart of the matter (problems within NASA; report by R. P. Feynman). *Scientific American* 255:62-4 Ag '86.

The explosion of the space shuttle Challenger has been examined in a report by a presidential commission and in a separate document by one of the commissioners, Richard P. Feynman of the California Institute of Technology. The immediate cause of the accident was a faulty seal in a joint between sections of the shuttle's right-hand solid-rocket booster. The Feynman report addresses underlying causes of the accident, pointing out that officials of Morton Thiokol, the booster's manufacturer had ignored their engineer's warnings regarding O-ring erosion. It also charges that NASA had systematically overrepresented its capabilities to Congress in order to receive funding. Although the presidential report is less critical of NASA, it does suggest the agency's initial plans for the shuttle were overambitious.

The Trouble with NASA

Hubble: heartbreak and hope. Abe Dane *Popular Mechanics* 167:130-1 O'90.

The Hubble Space Telescope, the largest and most expensive instrument ever put into space, is flawed by a mirror that has apparently been ground to the wrong shape. The technical name for the problem is spherical aberration, an elementary mistake in telescope optics. Its consequence is that light falling on the mirror cannot be sharply focused. The problem could have resulted from an error in the software instructions that guided the mirror's polishing and grinding devices, a fault with test equipment or human error. NASA has assembled an independent panel to investigate the problem, and it estimates that Hubble can perform about half of the work it was intended to accomplish.

Is NASA being militarized? James E. Oberg *Astronomy* 13:24+ F '85.

Widespread criticism of military involvement in the space shuttle program is groundless. The Defense Department was compelled to abandon its expendable-rocket programs and rely on the shuttle. The number of military payloads aboard the shuttle actually represents a decrease in military activity since the Pentagon's less-publicized rocket launches. Plans to build a shuttle base at Vandenberg Air Force Base stem from the unsuitability of Florida as a launch site for near-polar orbiting satellite missions. Such satellites are not overwhelmingly devoted to military applications. The militarization of NASA personnel is no more pronounced than during the Apollo program and serves important training functions. Moreover, the space shuttle is unsuitable for any space warfare applications. Unfortunately, factual refutations will not put an end to ideologically and politically motivated criticism.

NASA: what's needed to put it on its feet? Wayne Biddle *Discover* 8:30–4+ Ja '87.

A special report on NASA. NASA will need more funding and new ideas if it is to regain its old vigor in the wake of the Challenger disaster. Despite critics' claims that overreliance on the shuttle is what really hurt the space program, it remains NASA's top priority. This narrow focus on old technology, combined with limited funding, could stifle innovation, bankrupt the program's manufacturing base, and dull the appeal of a space career. The civilian space program is also threatened by the delegation of goal setting to an interagency body with conflicting interests, the Reagan administration's focus on military programs, and apparently unrealistic plans for a manned space station. Moreover, NASA does not have a strong leader who could sell ground-breaking ideas to Congress and the American people. The only innovative research that NASA will perform in the next few years will derive from projects begun long ago, like the Hubble Space Telescope and the Magellan probe.

The Americans in space. James M. Fallows *The New York Review of Books* 33:34+ D 18 '86.

The writer cites the space program as an example of why America has faltered in competition with the Soviet Union and countries with technically advanced market economies and discusses The Heavens and the Earth: A Political History of the Space Age, by Walter A. McDougall; Report to the President by the Presidential Commission on the Space Shuttle Challenger Accident, Vols. I-V; and The Aerospace Plane: Technological Feasibility and Policy Implications, by Stephen W. Korthals-Altes.

The brain drain. Brenda Forman *Ad Astra* 2:30–2 Ja '90.

The resignations of 27 senior managers from NASA in the spring of 1989 are likely to have serious consequences for the U.S. space program. The exodus was caused primarily by the defeat of a proposed pay raise and by the passage of new ethics legislation that limits the employment opportunities for top civil servants after they leave government. Hiring restrictions and no-growth budgets in the 1970s have reduced the pool of experienced NASA managers who can replace those who have left. The agency should establish communications links between new and former managers so that corporate memory can be preserved. To attract new talent to the space program and encourage young people to pursue engineering and science careers, NASA needs a budget that will support a solid base program.

From footprints to foothold. Robert G. Nichols *Astronomy* 17:48–53 Jl '89.

Part of a special issue celebrating the 20th anniversary of the *Apollo 11* moon landing. Twenty years after man first walked on the moon, there is talk of returning there to establish permanent lunar bases, which perhaps 20 years from now could be used to conduct long-term research and to launch commercial space industries. NASA's Office of Exploration, which was established in 1987 to determine America's long-term space objectives, has produced a scenario called Lunar Evolution. The program calls for a manned mission in 2004 to establish a permanent base camp on the moon, with supplies coming from a space station orbiting the earth. Eventually, a space station would be placed in lunar orbit, from which a cargo ship would transfer supplies to a lunar ferry traveling down to the base. Three major areas that would benefit from such a scenario are discussed: studying lunar geology, performing astronomy from the moon, and developing the moon industrially.

Twenty years after Apollo: is the U.S. lost in space? Stuart F. Brown *Popular Science* 235: 4, 63–75 Jl '89

A cover story discusses the future of the U.S. space program. The U.S. space program has been adrift since the *Apollo* moon landings, but there is now talk of a major new space effort. At the current space-program spending level, which is about 1 percent of the nation's budget, the United States could pursue unmanned interplanetary exploration and mount the Mission to Planet Earth project, a program of satellite monitoring. For about double the current space budget, the United States could establish a base on the moon or send people to Mars. The best scenario would involve a manned Mars mission in cooperation with the Soviet Union and other governments. Other articles present the opinions of space authorities on a variety of space-related issues.

U.S. must develop a solid base for future space exploration. John Yardley *Aviation Week & Space Technology* 129:105–6 N 28 '88

The president of McDonnell Douglas Aeronautics explains his proposal for the Foundation Program, which he believes would be a solid base for the future of the U.S. space program: The Foundation Program, composed of currently planned NASA projects, would ensure our space leadership in coming decades. Central to the program are the return to manned spaceflight, the deployment and operation of the permanently manned space station, the development of a robust space transportation capability, and the continued development of unmanned science and technology missions. The space station is particularly important to the task of American expansion into the solar system. From history, we have learned that we will see our space programs delayed and jeopardized unless we make both individual and financial commitments to leadership in space.

OUT OF THE CRADLE

Designing for the future. S. E. Sutphin *Space World* Y-4-292:
17-19+ Ap '88

NASA's new Advanced Design Program is giving thousands of undergraduate students the opportunity to participate in the space program of
the future. The program, managed by the Universities Space Research
Association [USRA], allows students to study advanced mission topics as
part of fully accredited courses. Every two years the USRA sends requests
for proposals to universities that have accredited engineering programs.
Universities whose proposals are selected are paired with a person at a
NASA center. Each university is given about $30,000, part of which goes
to hire a graduate teaching assistant, who acts as a liaison between the
school and the NASA contact person.

I hate Carl Sagan. Leonard David *Space World* Y-12-300: 4 D '88

Carl Sagan is the greatest popularizer of space exploration since Wernher
von Braun. One would hope that this position would be used to educate
the public as to what constitutes a mature, well-reasoned, and progressive
space program. Unfortunately, Sagan has come out in favor of a joint U.S.
-Soviet manned mission to Mars that puts symbolism over substance. He
believes that the Mars mission would promote world peace and that it
should be pursued at the expense of the rest of the U.S. space program,
including the lunar base and the space station. The scope of the U.S.
space program should not be narrowed, however. The Earth, Moon, and
Mars make up a triad of valuable worlds that are worth exploring and that
will support the future space voyages of people from many nations.

Predicting the space frontier. Tony Reichhardt *Space World* W-
10-274:8-13 O '86.

Laurel Wilkening, vice chairman of the National Commission on Space,
talks about the commission's recently published report and reaction to it.
She states her belief that the commission's 50-year program of exploring
and colonizing Earth orbit will eventually be realized, though not necessarily by the United States. She calls for long-term, multiyear procure-
ments for NASA but suggests that space is too big for one agency to
handle and that NASA should be broken up in a positive way.

Space cadets. Jeff Bloch *Forbes* 138: 78+ S 8 '86.

The U.S. Space Camp in Huntsville, Alabama, gives youths aged 11 to 16
a chance to learn about astronautics. Started in 1982 as an offshoot of the
Marshall Space Flight Center's space museum, the camp generates a type
of enthusiasm and optimism that NASA needs badly these days. Campers
wear replica astronaut flight suits, and some use equipment discarded
from actual NASA flights. Younger campers learn principles of rocketry
and propulsion and eventually launch 18-inch rockets carrying live crick-
ets. Older campers design space stations and prepare for a mock two-hour
space shuttle mission. Three-day adult sessions were such a hit last year
that they will be repeated this autumn.

The U.S. space station: a quarter-century of evolution. Philip D. Hattis *Technology Review* 91:28–40 J1 '88.

A U.S. space station was first proposed in the 1960s soon after humans were first launched safely into space. *Skylab*, a huge, fully equipped laboratory that was launched unmanned in 1973, was an early version of the space station. Skylab hosted three missions before its orbit decayed and it fell to Earth in 1979. From mid-1979 to the end of 1982, the Marshall Space Flight Center and the Johnson Space Center investigated space station designs that would utilize components transported by the space shuttle. Based on these and other studies, NASA decided on a multipurpose design for the station. In 1984, President Reagan endorsed the idea of a space station that would be operational within a decade. Political and technical hurdles remain before the space station can become a reality, but if Congress appropriates the requested funds and if shuttle operations resume as planned, the first components of a U.S. space station could be orbited in 1994 or 1995.